NEUROSCIENCE RESEARCH PROGRESS

SYMBIOTIC BIOFILMS AND BRAIN NEUROCHEMISTRY

NEUROSCIENCE RESEARCH PROGRESS

Additional books in this series can be found on Nova's website under the Series tab.

Additional E-books in this series can be found on Nova's website under the E-books tab.

NEUROSCIENCE RESEARCH PROGRESS

SYMBIOTIC BIOFILMS AND BRAIN NEUROCHEMISTRY

**ALEXANDER V. OLESKIN
VLADIMIR A. SHISHOV
AND
KRISTINA D. MALIKINA**

Nova Biomedical Books
New York

Copyright © 2010 by Nova Science Publishers, Inc.

All rights reserved. No part of this book may be reproduced, stored in a retrieval system or transmitted in any form or by any means: electronic, electrostatic, magnetic, tape, mechanical photocopying, recording or otherwise without the written permission of the Publisher.

For permission to use material from this book please contact us:
Telephone 631-231-7269; Fax 631-231-8175
Web Site: http://www.novapublishers.com

NOTICE TO THE READER

The Publisher has taken reasonable care in the preparation of this book, but makes no expressed or implied warranty of any kind and assumes no responsibility for any errors or omissions. No liability is assumed for incidental or consequential damages in connection with or arising out of information contained in this book. The Publisher shall not be liable for any special, consequential, or exemplary damages resulting, in whole or in part, from the readers' use of, or reliance upon, this material.

Independent verification should be sought for any data, advice or recommendations contained in this book. In addition, no responsibility is assumed by the publisher for any injury and/or damage to persons or property arising from any methods, products, instructions, ideas or otherwise contained in this publication.

This publication is designed to provide accurate and authoritative information with regard to the subject matter covered herein. It is sold with the clear understanding that the Publisher is not engaged in rendering legal or any other professional services. If legal or any other expert assistance is required, the services of a competent person should be sought. FROM A DECLARATION OF PARTICIPANTS JOINTLY ADOPTED BY A COMMITTEE OF THE AMERICAN BAR ASSOCIATION AND A COMMITTEE OF PUBLISHERS.

Additional color graphics may be available in the e-book version of this book.

Library of Congress Cataloging-in-Publication Data

Oleskin, A. V. (Aleksandr Vladimirovich)
Symbiotic biofilms and brain neurochemistry / Alexander V. Oleskin, Vladimir I. Shishov, and Kristina Malikina.
p. ; cm.
Includes bibliographical references and index.
ISBN 978-1-61761-005-9 (softcover)
1. Biofilms. 2. Neurochemistry. 3. Gastrointestinal system--Microbiology. I. Shishov, Vladimir I. II. Malikina, Kristina. III. Title.
[DNLM: 1. Biofilms--growth & development. 2. Bacterial Physiological Phenomena. 3. Neurotransmitter Agents. 4. Signal Transduction. QW 90]
QR100.8.B55O44 2010
579'.17--dc22

2010027252

Published by Nova Science Publishers, Inc. † *New York*

Contents

Preface		vii
Chapter I	Introduction	1
Chapter II	Basic Properties of Microbial Biofilms	3
Chapter III	Quorum Sensing and Microbial Biofilms	11
Chapter IV	Symbiotic Microflora and Its Biofilms	15
Chapter V	Influence of Neurotransmitters on GI Microflora and Its Biofilms	19
Chapter VI	Neurotransmitter Biosynthesis and Release by Microorganisms: Implications for Symbiotic Biofilms	33
Chapter VII		39
Conclusion		39
References		41
Index		51

Preface

An overwhelming majority of known species of microorganisms form biofilms, i.e. spatially and metabolically structured communities embedded in the extracellular biopolymer matrix. Biofilm development is a complex multi-stage process involving reversible and, at a later stage, irreversible attachment of microbial cells to the substrate surface, matrix formation, three-dimensional structuring of the whole community including the formation of mushroom- or pillar-shaped structures and, finally, the degradation of the biofilm and the dispersion of the cells involved. These processes are considered in the example of microorganisms that interact with the animal or human organism, playing the roles of symbionts or pathogens. In particular, the microorganisms of the gastro-intestinal (GI) tract interconvert between two different lifestyles: they can exist as planktonic cells in the intestinal lumen or form part of a biofilm attached to the mucous membrane of the GI tract. The GI microflora, including the biofilm-forming cells, is subject to regulation by the metabolites and chemical signals produced by the human/animal host.

The data published in the literature and our own findings suggest an important role of host-produced neuromediators, such as amines, peptides, and nitric oxide, which regulate biofilm formation by influencing microbial growth rate, the aggregation of microbial cells, the formation of microcolonies, and matrix synthesis. Our results were obtained using high efficiency liquid chromatography and revealed that cells of various symbiotic and pathogenic bacterial species contain serotonin, norepinephrine, and dopamine, as well as their precursors and oxidative deamination products. It follows from our studies with *Escherichia coli* that the culture of this symbiotic, biofilm-forming bacterium releases amine neuromediators and their precursors/products into the culture fluid in concentrations of 10–100 nM, which are sufficiently high enough to cause the host's physiological response. These facts and other relevant data are considered in the article in terms (i) of the autoregulatory role of neuromediators in the biofilm-forming microbial population and (ii) the microbially-produced neuromediator amines' impact on the human organism. Of particular interest in this respect are the data that the culture fluid of *E. coli* contains over 1 µM DOPA, the catecholamine precursor. DOPA crosses the gut-blood and blood-brain barriers. In the brain, DOPA is converted to dopamine and thereupon to norepinephrine that regulate brain processes involved in locomotion, affection, sociable and dominant behavior, as well as aggression.

Chapter I

Introduction

A human's brain is beyond any doubt the central regulator of his social behavior and the material substratum of his psyche. The brain cannot function without neurotransmitters, i.e. substances that transmit information between nervous cells (neurons) across the synaptic cleft separating them (or between a neuron and a muscle/gland cell that caries out brain signals). A sufficiently large number of neurotransmitters, including biogenic amines (serotonin, histamine, dopamine, norepinephrine, etc.) are multifunctional agents – they combine the functions of neurotransmitters with those of hormones or local tissue factors (histohormones). Another important point is that many neurotransmitters are highly conserved compounds that perform communicative and regulatory functions in representatives of various types of living organisms including animals, plants, fungi, and unicellular creatures.

This article is concerned with the functional roles performed by neurotransmitter compounds in the world of microorganisms. Particular emphasis will be placed on the involvement of these compounds in the development and functioning of biofilms formed by microflora inhabiting a human/animal organism. Special attention is to be paid to the role of neurotransmitters in the "dialog" among various microbial species within biofilms and between symbiotic/parasitic biofilms and the host organism as well as the influence of this "dialog" upon the host's neurochemistry, which, in turn, influences his health, psyche and social behavior.

This subject is of relevance to the novel interdisciplinary field of science referred to as **biopolitics**. Presently, this field is actively developing around the world; it embraces the totality of social and political implications of the life sciences (see Somit and Peterson, 1998; Masters, 2001; Oleskin, 2007).

One important subfield of biopolitics deals with research on the biological basis of human political behavior considered a result of the evolution of human ancestors over the course of millions of years. This subfield of biopolitics is closely related to neurophysiology and gives special attention to the importance of neurotransmitters (that convey impulses between nervous cells) for the social behavior of animals and humans.

This contribution concentrates upon **biofilms,** i.e. "multicellular, matrix-enclosed assemblies... that are found throughout the biological world" (Romeo, 2006, p.7325). Special emphasis will be placed upon bacteria that inhabit various niches of the human organism, influence human neurochemistry and behavior, and are influenced, in their turn, by the human organism producing its own regulatory substances including neurotransmitters. Hence this work deals with what could be referred to as "microbial biopolitics".

Chapter II

Basic Properties of Microbial Biofilms

An overwhelming majority of known bacterial species can form biofilms, "matrix-enclosed microbial accretions that adhere to biological or non-biological surfaces" (Hall-Stoodley et al., 2004, p.95). The diversity of biofilms is fascinating: they range "from patchy monolayers on some surfaces through very thick gelatinous masses associated with cooling systems to filamentous accretions near sewage outlets" (Wimpenny et al., 2000, p.662). Biofilms can be formed by cells of a single bacterial species (and even single-species films are heterogeneous because they include cells with different phenotypes) or represent multispecies systems. Some biofilms include representatives of different kingdoms. For instance, the films of a methanogenic association are composed of cells of eubacteria and archaea.

Biofilms are of considerable theoretical interest because they are advanced microbial social systems, "cities of microbes" (Watnick and Kolter, 2000) that are to some extent similar to multicellular organisms. In particular, some microbial biofilms are characterized by the distribution of functions among their cells, an effective coordination of cell behavior, a unitary life-cycle (ontogeny) of the whole biofilm, and even the capacity to regenerate after an injury (figure 1), as demonstrated in studies with cyanobacterial biofilms (Sumina, 2006).

Biofilms hold much practical value. A large number of important biotechnological developments involve microbial biofilms exemplified by the traditional French technique of producing vinegar with the help of *Acetomonas* biofilms on wood chips. The "tea fungus" (the medusomycete, or the Kombucha, see Yurkevich and Kutyshenko, 2002), the producer of a useful medicinal drink, represents a thick multispecies biofilm consisting of bacterial and yeast cells. Biofilms normally overgrow the mucous membrane of the large intestine of animals and humans and is considered a special "extracorporal organ" of their organism that carries out a large number of useful functions (see below for more details).

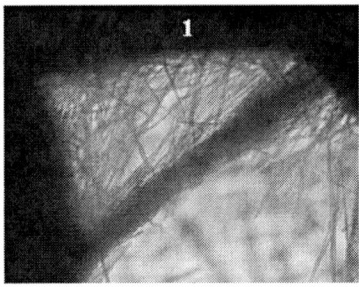 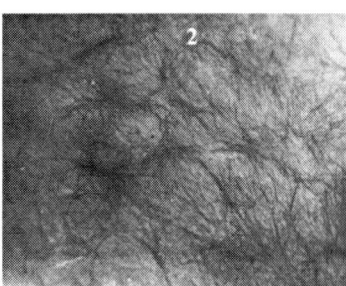

Figure 1. The biofilm of the cyanobacterium *Oscillatoria terebriformis* regenerates after being ruptured with a needle (1) and spreading $CaCO_3$ powder over its surface (2). *O. terebriformis* filaments grow across the "wound" on the biofilm surface in (1) and gradually overgrow the $CaCO_2$ layer in (2). Magnification, 250. The pictures are a generous gift from Dr. Eugenia L. Sumina.

However, microbial biofilms can also do much harm. They can cause the destruction of various materials and constructions (biofouling). A serious threat is posed by the biofilms of pathogenic microorganisms. "Biofilm formation is an important aspect of many, if not most, bacterial diseases, including native valve endocarditis, osteomyelitis, dental caries, middle ear infections, ocular implant infections, and chronic lung infections in cystic fibrosis patients," (Jefferson, 2004, p.163).

Despite their diversity, most microbial biofilms possess the following typical properties (reviewed, Stoodley et al., 2002; Lemon et al., 2008; Senadheera and Cvitkovitch, 2008; Karatan and Watnick, 2009):

- **Spatial organization,** i.e. the formation of two- and three-dimensional structures in a biofilm, exemplified by local cell aggregates (microcolonies), cavities (pores and channels), lipid membrane vesicles, the outer cover of the biofilm including the biofilm-coating lipid bilayer (Tetz et al., 1992, 2004), and the biofilm's functional "organs" such as the O_2-transferring hemosomes of *Alcaligenes* (Duda et al., 1995, 1998) and fruiting

bodies with maturating spores (in myxobacteria) or their analogs (in bacilli)
- **Metabolic organization** implying a directed metabolite flow in a biofilm
- **The intercellular biopolymer matrix** that is responsible for the adherence of bacterial cells to the substratum and their cohesion; the matrix is composed of polysaccharides (Sutherland, 2001), e. g., poly-β-1,6-N-acetylglucosamine (PNAG), colanic acid (in the matrix of *Escherichia coli* biofilms), glycoproteins, polypeptides exemplified by polyglutamic acid (Safronova and Botvinko, 1998), and extracellular DNA molecules. The matrix is involved in maintaining the structural integrity of a biofilm, protecting microbial cells from deleterious environmental factors, masking the cells' surface antigens to prevent their recognition by host immune cells, and creating a hydrophilic environment to promote the spreading of metabolites and signal molecules within the biofilm
- **Adherence to a phase boundary** such as solid body/liquid, solid body/air, liquid/air, or liquid/liquid boundaries (reviewed, Nikolaev and Plakunov, 2007).

The typical structure of a biofilm is formed stepwise (figure 2). Initially, a **transient attachment** of microbial cells **(primary colonizers)** occurs, which is due to their interactions with the substratum involving flagella, pili, fimbria, and the proteins of the outer membrane of gram-negative bacteria. This stage is followed by the **permanent attachment** of microbial cells to the surface. For example, motile bacterial cells first attach one of their poles by means of their flagella to a substratum; thereupon, one of their sides contacts the surface and is anchored there. Subsequently, microbial cells **spread** on the substratum colonized by motile bacterial cells. This is accompanied by the formation of local cell aggregates and the intracellular matrix with characteristic cavities and the biofilm cover including a part of the matrix as well as lipid bilayer structures (Tetz et al., 1993, 2004; Pavlova et al., 2007). The development of a majority of biofilms includes the stage characterized by the **attachment of new microbial cells (secondary colonizers)** to the substratum-anchored cells, which results in the formation of multilayer biofilms. Eventually, a single- or multiple-species biofilm with a developed structure is formed; this biofilm can display a lamellar structural pattern or contain mushroom- or pillar-shaped formations.

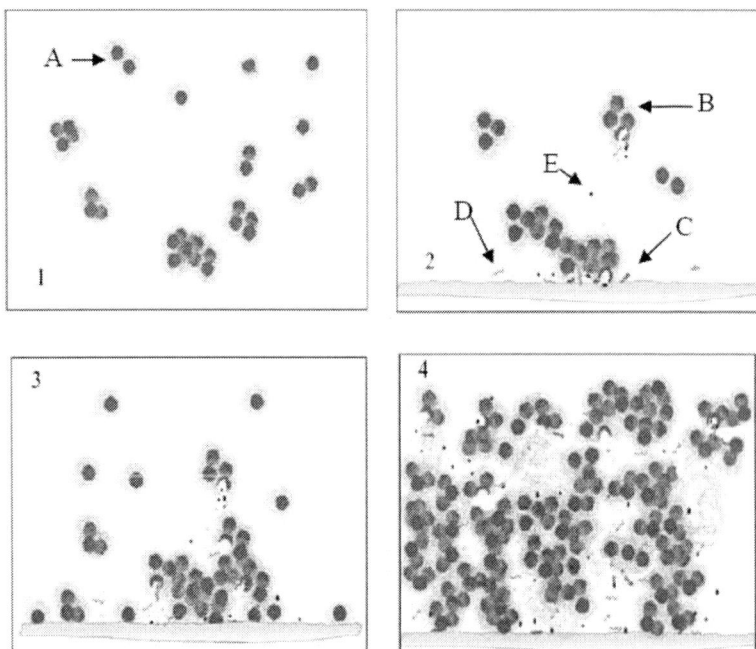

Figure 2. Stages of biofilm formation by *Staphylococcus epidermidis* 33 in the human oral cavity (a scheme). The Figure demonstrates the consecutive stages of the transition from a planktonic lifestyle (1) via the attachment of primary colonizers (2) and extracellular matxix synthesis (2, 3) to the formation of three-dimensional pillar- and mushroom-like structures (4). A, *S. epidermidis* cells; B, matrix elements; C, extracellular DNA; D, extracellular RNA; E, proteins and peptides including enzymes and quorum-sensing autoinducers. The pictures are a generous gift from Dr. Vladimir P. Korobov.

The final stage of a biofilm's life-cycle is its **dispersal**; its microbial cells return to the planktonic mode of existence. This involves the disruption of the cell attachment to the substratum, which is caused by the synthesis of surfactants and enzymes that degrade the matrix including its components (adhesins) directly responsible for cell attachment to the substratum and to one another. For instance, the oral cavity-inhabiting bacterium *Actinobacillus actinomycetemcomitans* produces an enzyme that degrades the adhesin PNAG (Itoh et al., 2005). Biofilm dispersal also involves the suppression of *de novo* adhesin synthesis. Detailed studies conducted with the opportunistic pathogen *Pseudomonas aeruginosa* that may inhabit various niches in the human organism have demonstrated that biofilm dispersal involves an endogenous prophage and the death of a part of the biofilm cells associated with the transition of the remaining viable cells to the planktonic mode of existence. As a result, the surface-adherent microcolonies of biofilms undergo disintegration, converting into hollow shell-like structures. The whole phenomenon is referred to as "seeding dispersal" (Romeo, 2006).

Owing itself to the diversity of biofilm types (the same species can form structurally different biofilms, depending on its cultivation conditions and the genetic peculiarities of the given strain), different variants of the above stages can occur. Moreover, the biofilm life-cycle may lack some of these stages. For instance, some bacteria form only matrix-embedded monolayer biofilms, i.e. no attachment of secondary colonizers to the surface-adherent primary colonizers occurs (Karatan and Watnick, 2009). Multilayer biofilms (mats) formed by associations of photosynthetic or sulphate-reducing microorganisms are often characterized by a lamellar structure lacking the mushroom- or pillar-like formations that are peculiar to a large number of other biofilms, e.g. those formed by oral streptococci.

The biofilm formation process is influenced by complex environmental factors as well as by regulatory agents formed by microbial cells per se. Cultivation conditions (pH, temperature, nutrient substrate concentration, pO_2, osmolarity, surface hydrophilicity/hydrophobicity degree, shear force, etc.) produce their effects on microbial biofilms. For instance, nutrient limitation results in enhanced biofilm formation by *Salmonella enteric* var. Typhimurium, which involves the operation of the stationary-phase σ factor RpoS (Gerstel and Romling, 2003). In contrast, the development of *Vibrio cholerae* biofilms is enhanced in a nutrient-rich medium, and RpoS represses the genes involved in biofilm formation (Yildiz et al., 2004). It was established that biofilm formation in pathogenic and non-pathogenic *E. coli* strains is conditional upon their cultivation conditions including medium composition. Most tested *E. coli* strains failed to form biofilms on the rich Luria-Bertani medium but formed them on a minimal medium and on diluted porcine intestinal mucus (Reisner et al., 2006).

In a large number of studies, the data obtained suggested that biofilm formation is promoted by factors that cause stress in microorganisms. This is exemplified by the fact that the addition of subinhibitory concentrations of antibiotics, such as the aminoglycoside tobramycin, induces biofilm formation in pathogenic *E. coli* strains (Hoffman et al., 2005). Biofilm formations can be regarded, therefore, as a **stress response**. In particular, the biofilm-coating membrane, whose rigidity is enhanced by the presence of such lipids as cardiolipin, is, in all likelihood, functioning as a protective barrier between the bacterial association and the environment (Tetz et al., 2004). However, microorganisms have other reasons for forming biofilms, including the following incentives (Jefferson, 2004):

- **Sequestration to a nutrient-rich medium,** colonization of a favorable ecological niche. This is exemplified by biofilm formation in the gastro-intestinal (GI) tract; importantly, different species and strains, e. g. pathogenic and non-pathogenic *E. coli* strains, can compete for the resources available in various zones of the GI tract (Bansal et al., 2007).
- **Utilization of the benefits of a social lifestyle** including "labor distribution" in respect to metabolic activities. For example, a methane-producing association includes bacteria with hydrolytic activities that convert the organic substrate into monomers. They are made available for acido- and acetogenic microflora, which, in turn, provides substates to be utilized by methane-producing archeae. The advantages of biofilms as social systems also include "protection against... predation by protozoa," and "enhanced ability to transfer genetic information," (Romeo, 2006, p.7325).
- **Biofilms as the default mode of existence**; as the normal lifestyle of most microorganisms. The existence of microbial cells in suspensions (the planktonic lifestyle) represents, in these terms, a temporary adaptation aimed at searching for a new suitable habitat for biofilm formation or just an *in vitro* artifact (Jefferson, 2004).

Regardless of the incentive involved, biofilm development and dispersal are subject to control by a complex of intra- and intercellular regulators. Microorganisms have special gene blocks involved in the planktonic cells–biofilm interconversion, including genes responsible for the adherence of microbial cells to substrata and to other cells, such as the *algC* gene required for the synthesis of alginate, a matrix component in *P. aeruginosa,* and *wcaB* gene involved in colanic acid synthesis in *E. coli.* Biofilm formation in *E. coli* implicates the expression of genes that are involved in the production of bacterial cell surface structures, such as the *csgA* gene required for the formation of curli fibers (reviewed, Jefferson, 2004). Paramount to biofilm formation are quorum sensing systems, which will be considered in some detail below.

Gene expression during biofilm formation is influenced by a variety of intracellular regulatory factors. Important functions are performed, particularly in gram-negative bacteria, by cyclic diguanylate monophosphate (c-di-GMP). Accordingly, the regulation of the activities of the c-di-GMP-synthesizing enzyme diguanylate cyclase (DGC) and of the c-di-GMP-degrading phosphodiesterase A (PDEA) are essential for the operation of the intracellular network of regulatory agents involved in biofilm formation/dispersal. "It is now widely accepted that cyclic di-GMP (c-di-GMP) signaling, first described to control extracellular cellulose biosynthesis in *Gluconacetobacter xylinus*..., is involved in the modulation of matrix components, control of autoaggregation of planktonic cells, and biofilm formation in several microorganisms... High c-di-GMP concentrations have been shown in *Salmonella enterica* serovar Typhimurium to stimulate biofilm formation and EPS[1] production (and thus, adhesiveness) but to suppress motility, while low concentrations inhibited biofilm formation, repressed the production of EPS, and stimulated swimming and swarming motilities," (Morgan et al., 2006, p.7335). c-di-GMP operates as a multilevel intracellular regulator that influences transcription, translation, and posttranslational protein modification and activity modulation. For example, c-di-GMP binds to the enzyme PelD and regulates the synthesis of the matrix polysaccharide PEL (Lee et al., 2007a).

The intracellular c-di-GMP concentration is decreased in response to environmental stimuli such as sudden changes in the nutrient concentration (in *P. aeruginosa*, this can be an increase in glutamate concentration) and oxygen depletion in the interior of the biofilm. The decrease in the c-di-GMP concentration results in a large number of bacterial species in biofilm dispersal (reviewed, Camilli and Bassler, 2006; Karatan and Watnick, 2009). The influence of environmental factors on the c-di-GMP pool is mediated by the chemotaxis protein BdlA that contains two PAS (Per-Arnt-Sint) domains involved in receiving a variety of extracellular signals including nitric oxide (Barraud et al., 2009a) that will be considered in more detail below.

---

[1] EPS is extracellular polysaccharide.

Chapter III

Quorum Sensing and Microbial Biofilms

On the population level, biofilm formation is regulated by signal molecules (autoinducers). Special emphasis should be placed on **quorum-sensing systems**. These systems regulate processes carried out by bacterial cells and structures formed by them including biofilms, taking account of the density of a microbial population. An increase in population density results in an increase in the autoinducer concentration formed by microbial cells. The autoinducer binds to the proteinaceous response regulator[2]. The resulting complex binds to specific operons and stimulates or suppresses their expression. Quorum-sensing systems of most gram-negative bacteria use N-acetylated homoserine lactones (N-AHLs), referred to as autoinducers-1, or AI-1, in related literature.

A classic system is responsible for the bioluminescence of the marine bacterium *Vibrio fischeri* (Fuqua et al., 1994). The bioluminescence occurs only in concentrated bacterial populations growing in the light organ of the mollusc *Euprymna scolopes*, where the population density reaches $10^{10}-10^{11}$ cells per ml. The quorum-sensing system consists of two main gene blocks. One of them is the *luxICDABEG* operon. It includes the *luxI* gene encoding the protein responsible for the synthesis of the autoinducer N-(3-oxohexanoyl)-L-homoserine lactone, while the other genes in this operon encode the components of the enzyme complex that catalyzes

---

[2] It should be noted that the autoinducer concentration may increase up to the threshold level because of decreasing the volume of the compartment containing the microbial population, not increasing the population density.

bioluminescence. The second operon includes the *luxR* gene whose protein product binds the autoinducer, provided that its concentration produced by *V. fischeri* cells and, therefore, the population density is above the threshold level. The autoinducer–LuxR complex binds to the promoter of the *luxICDABEG* operon and activates its transcription. A large number of other quorum-sensing systems in gram-negative bacteria are based on the same principle, despite minor differences. The autoinducer is produced by an N-AHL synthase of the LuxI family of proteins and is recognized by an N-AHL receptor protein of the LuxR family.

N-AHL-dependent quorum-sensing systems are involved in regulating biofilm formation, as exemplified by *P. aeruginosa*. This bacterium possesses two systems of this type, referred to as *lasI-lasR* and *rhlI-rhlR*. A mutant with an impaired *lasI-lasR* system as well as a double mutant lacking both systems formed abnormal thin flat biofilms that were destroyed by the detergent sodium dodecyl sulfate (SDS). The wild-type strain and a mutant with an impaired *rhlI-rhlR* system formed SDS-resistant three-dimensional biofilms with mushroom- or pillar-like structures separated by water-filled cavities. From this data it was evident that only the *lasI-lasR* quorum-sensing system is essential for biofilm formation in *P. aeruginosa* (Davies et al., 1998).

Gram-positive bacteria is characterized by quorum-sensing systems with peptide autoinducers that are perceived by two-component signal transduction systems (including a sensor and a response regulator) that activate the transcription of target genes. The potential pathogen *Staphylococcus aureus* uses a peptide with a thiolactone ring (AIP) as the autoinducer. This quorum-sensing system negatively regulates biofilm formation because AIP bound to the AgrC protein (the response regulator) activates the transcription of RNA III that codes for the hemolysin protein with surfactant properties and, in addition, stimulates the expression of the extracellular proteases Aur and Spl involved in degrading biofilms (reviewed, Karatan and Watnick, 2009).

Widespread in the microbial world is quorum-sensing autoinducer AI-2 implicated in regulating biofilm formation in a number of gram-positive and -negative bacteria and formed with the help of the enzyme LuxS. AI-2 is 2-methyl-2,3,3,4-tetrahydroxytetrahydrofuran (THMF), characterized by either the 2S, 4S- or the 2R, 4S-conformation and forming a boron diester in many bacterial species. It plays a somewhat controversial role in microbial cells because it serves both as a sink for the toxic metabolite S-adenosylhomocysteine and as an interspecies autoinducer (reviewed, Ryan and Dow, 2008). The autoinducer role is illustrated by the fact that bacteria which do not produce AI-2, such as *P. aeruginosa*, can respond to it nevertheless. The virulence of this pathogen and biofilm formation in human lungs are stimulated by AI-2 produced by normal oropharyngeal microflora (Duan et al., 2003). Biofilm formation by the pathogenic enterohemorrhagic strain of *E. coli* (EHEC) causing bloody diarrhea and blood poisoning, is stimulated by AI-2 (Gonzalez Barrios et al., 2006).

This example provides a link to the subject of the final sections of this contribution that deal with the role of signal molecules (primarily, of substances performing neurotransmitter functions in the nervous system) in biofilm formation by the symbiotic microflora of humans and animals. In this context, quorum-sensing autoinducer AI-3, previously unmentioned, will be considered.

Chapter IV

Symbiotic Microflora and Its Biofilms

Symbiotic microorganisms, most of whom are capable of forming biofilms, inhabit various niches on the surface and inside the animal or human body. The total microbial population of an adult human counts ca. 10^{14} cells, which is about an order of magnitude larger than the total number of human cells (Berg, 1996). Microorganisms grow on the skin (and their maximum concentrations are detected on the skin between fingers, on the foot soles, in the inguinal folds and the armpits, on the scalp, and around the breasts), on the eye conjunctiva, and on the mucosa of the upper respiratory tract and the urogenital system. However, this work emphasizes the role of microorganisms and their biofilms in the gastrointestinal (GI) tract and especially in the large intestine.

Human GI microflora includes representatives of over 50 genera and between 400 and 1000 species of microorganisms (Xu et al., 2003; Valyshev and Gilmutdinov, 2006). GI microorganisms can exist as planktonic cells in the intestinal lumen or as biofilms in the mucus layer overlying the epithelium, mucus within intestinal crypts, and the surface of mucosal epithelial cells (Kaper, Sperandio, 2005). Normally, the microflora performs a number of vital functions: it is involved in food digestion, intestinal motility regulation, the maintenance of the acidity (pH), temperature, and gas phase composition within normal limits, promotes the maturation of the intestinal mucosa and contributes to its barrier function with respect to harmful substances and pathogenic microorganisms and viruses, stimulates the activity of the immune system, and provides the organism with useful compounds such as vitamins, short-chain fatty acids, and neuromediators (reviewed, Valyshev and Gilmutdinov, 2006). On the whole, the microflora of the intestinal lumen and mucosal biofilm is a special multifunctional "microbial organ" which is similar to the liver in its weight (1 - 1.5 kg on average).

Nevertheless, the Russian microbiologist and immunologist Elia I. Metchnikov asserted that human normal microflora is not necessarily optimal. Microflora can cause serious problems (**dysbacterioses**) resulting from negative changes in its qualitative and quantitative composition, such as a decrease in the concentrations of useful bacteria (bifidobacteria, lactobacilli, bacteroids, and nonpathogenic *E. coli*) frequently associated with an increase in potential pathogens including *Clostridium difficile* causing potentially lethal pseudomembraneous colitis, the pathogenic yeast *Candida albicans,* and purulent flora (reviewed, Valyshev and Gilmutdinova, 2006; Bai and Ouang, 2006). The currently increasing frequency of dysbacterioses is undoubtedly due to the frequent employment of antibiotic and hormone preparations, serious environmental problems such as the accumulation of various pollutants and especially radioactive waste, malnutrition, and stress.

Use of dysbacterioses have resulted in serious problems including not only intestinal diseases (irritated bowel syndrome, Crohn's disease, ulcerous colitis, and colon cancer), but also liver dystrophy, rheumatoid arthritis, spondylosis, multiple organ failure and mental disorders (autism, Tourette's syndrome, ADHD, etc.). The influence of microflora on our physical and mental health is mediated by chemical agents including neuromediators.

Human microflora is responsive to alterations in the host's neurophysiological and psychological state. For example, anger and fear cause a transient increase in the number of *Bacteroides fragilis* subsp. *theaiotaomicron* cells within the bacterial population of human feces (Hawrelak, 2008). Characteristic bacterial species such as *Anaerofustis stercohominis, Anaerotruncus colihominis, Clostridium bolteae,* and *Cetobacterium someria* accumulate in a child's excrement at a late stage of autism (Valyshev and Gilmutdinova, 2006).

There is evidence that social stress changes the qualitative and quantitative composition of the microflora of the people involved. Studies on stress caused by separating young rhesus macaques from their mothers revealed that maternal separation resulted in statistically significant changes in fecal microflora. The number of lactobacilli decreased in their feces, starting from day 2 of separation, while the concentration of pathogens (*Shigella spp.* and *Campylobacter spp.*) increased (Hawrelak, 2008). Accumulation of potential pathogens with a decrease in the number of useful bifidobacteria and lactobacilli was also observed in Russian cosmonauts after a long space flight, which was probably due to stress, despite the possible contribution of a low-fiber diet (Valyshev and Gilmutdinova, 2006). In turn, microbes can influence the host organism's state. Some of them can relieve stress, and this is one of the functions of useful microbes such as bifidobacteria and lactobacilli. They have been traditionally used to prepare fermented milk products such as yogurt. Presently, these useful bacteria form part of special drug preparations (probiotics) that improve the health and the psychological state of patients by overcoming dysbacterioses and introducing normal microflora into the intestines.

However, Metchnikoff's idea that normal microflora is not optimal should be re-emphasized at this point. Some components of our microflora can aggravate stress by causing an infectious disease. For instance, when the pathogen *Helicobacter pylori* is activated, it can cause an ulcer if the individual sustains severe stress (Murrison, 2001).

The fact that GI microorganisms are characterized by high density and diversity suggests intense communication within this community and between it and the host to coordinate any number of advantageous processes (Clarke et al., 2006). Discussion has already started on communication facilities inside the microbial symbiotic biofilm due to the consideration of the influence of quorum-sensing systems on biofilm formation.

The issue to be discussed in the further sections of this article is the impact of neurotransmitters, a group of substances produced both by the host organism and by its microflora, on GI microorganisms and their biofilms.

Chapter V

Influence of Neurotransmitters on GI Microflora and Its Biofilms

Neurotransmitters include several groups of substances such as (i) biogenic amines comprising catecholamines (dopamine, norepinephrine, and epinephrine[3]), serotonin, histamine, etc.; (ii) amino acids such as aspartic, glutamic, and γ-aminobutyric acid, glycine, and β-alanine; (iii) peptides including endorphins, enkephalins, dinorphins, and substance P; and (iv) gaseous neurotransmitters: nitric oxide, carbon monoxide, and sulfur hydrogen. This paper will discuss only the neurtransmitters that have been investigated in studies with microbial systems (particularly biofilms).

1. Catecholamines derive from the amino acid tyrosine whose hydroxylation yields dioxyphenylalanine (DOPA), the immediate catecholamine precursor. Dopamine and norepinephrine are important neurotransmitters. Norepineprine activates the brain, increases locomotive activity, decreases the anxiety level, and promotes human/animal aggressive behavior (Dubynin et al., 2003). As for dopamine, it is also involved in maintaining a sufficiently high level of locomotive activity. Dopamine also enables an individual to perform complex voluntary movements while suppressing involuntary ones and maintain an active wakeful state. Dopamine promotes pleasure-seeking behavior (hedonic behavior, Berridge and Robinson, 1998).

---

[3] Epinephrine mainly performs the function of an adrenal hormone, whereas chemically similar norepinephrine functions both as a hormone and a major neuro transmitter

Of particular interest are microbiological aspects of the functions of catecholamine neurotransmitters. The response to stress involves releasing catecholamines (epinephrine, norepinephrine, and, to a lesser extent, dopamine) into the bloodstream. Studies with mice revealed that the release of norepinephrine into the bloodstream caused by damaging noradrenergic nervous cells with a neurotoxin drastically increased the number of bacterial cells in the cecal wall and lumen, with *E. coli* as the predominant species (Lyte, Bailey, 1997). Norepinephrine and other catecholamines can exert an indirect influence on the microflora by suppressing immunoglobulin A synthesis and/or release. They also stimulate intestinal motility and bile flow, which promotes the growth of such bacteria as *Bacteroides*. This apparently accounts for the above data on an increased *Bacteroides* content in the feces of angry or frightened individuals.

Importantly, catecholamines also produce a direct stimulatory effect on the growth of human organism-inhabiting microorganisms, including such pathogens as *Yersinia enterocolitica* causing intestinal inflammatory diseases; virulent *E. coli* strains that are responsible for intestinal infections and other serious problems including blood poisoning, *Shigella* causing dysentery, *Salmonella* responsible for food-caused infections, *Pseudomonas aeruginosa* involved in purulent inflammation in various organs (Lyte, Ernst, 1993; Freestone et al., 1999, 2007a), *Bordetella pertussis* and *B. broncioseptica* responsible for inflammatory diseases involving respiratory organs (Freestone, Lyte, 2008), *Aeromonas hydrophila* (Kinney et al., 1999), and *Staphylococcus epidermidis* (Lyte et al., 2003).

Based on the data obtained in our lab (Anuchin et al., 2008), catecholamines stimulate biomass accumulation (estimated nephelometrically) and cell proliferation (determined from colony-forming unit number increase) in nonpathogenic *E. coli* K-12 (strain MC4100). They are also stimulatory to a number of microorganisms that do not inhabit the GI tract including the eukaryote *Saccharomyces cerevisiae* (Malikina et al., 2010), which, nevertheless, can be used as a probiotic (Martins et al., 2005).

Catecholamines promote the adherence of GI microflora to intestinal mucosa and biofilm formation on it. A significant stimulation of biofilm formation as visualized by scanning electron microscopy was attained after the addition of catecholamines to the skin-inhabiting bacterium *S. epidermidis* (Lyte et al., 2003) that also occurs in the GI tract. Apart from cell proliferation, catecholamines stimulate the synthesis of adhesins, toxins, and other virulence factors (Lyte et al., 1996; Clarke et al., 2006). Norepinephrine increases the adhesion of pathogenic enterohaemorrhagic *E. coli* (EHEC) to cecal mucosa (Chen et al., 2003), colonic mucosa (Green et al., 2004), and ileum (Vlisidou et al., 2004). Catecholamines also stimulate the flagellar motility of EHEC (Sperandio et al., 2003).

The difference between the effects produced by catecholamines on pathogenic and nonpathogenic microflora should be emphasized. The growth of pathogenic enterobacteria is maximally stimulated by norepinephrine, whose release into the bloodstream is characteristic of a macroorganism's response to serious infection. Dopamine, the minor component of the response, is active at comparatively high concentrations. For instance, norepinephrine and dopamine at concentrations of 3 and 100 µM, respectively, stimulated the growth of the pathogenic strain *E. coli* JPN10 (O44:H18) on the liquid SAPI medium with blood serum (Burton et al., 2002). However, in the nonpathogenic symbiotrophic strain *E. coli* K-12 MC 4100, dopamine stimulated proliferative activity and biomass accumulation to a greater extent than norepinephrine (Anuchin et al., 2008).

The formation of local cell aggregates developing into microcolonies is the initial stage of a biofilm's life cycle (see above). In our studies, dopamine and norepinephrine produced different effects on this process in *E. coli* K-12 populations growing on LB agar. The catecholamines were added upon inoculation. We counted solitary cells and compact cell groups and calculated the ratio between them. From our data we concluded that dopamine inhibited cell aggregation, whereas norepinephrine stimulated it (Anuchin et al., 2008).

Dopamine stimulated cell proliferation in the eukaryote *Saccharomyces cerevisiae*, while norepinephrine did not significantly influence this process (Malikina et al., 2010).

Taking into account the difference between the effects of two chemically similar compounds, dopamine and norepinephrine (which only contains an extra OH group in the side chain) our research suggests that these effects at least partly result from specific interactions involving receptors on the surface of microbial cells that individually recognize neurotransmitters. We demonstrated in our studies with *S. cerevisiae* that the apomorphine that binds to the dopamine-specific D_1 and D_2 receptors stimulates cell proliferation like dopamine (Malikina et al., unpublished).

The effects of catecholamines on microorganisms and their biofilms are interpreted in the literature in terms of quorum-sensing communication in microbial populations (Clarke et al., 2006; Bansal et al., 2007). The influence of quorum-sensing systems involving autoinducers AI-1 (N-AHLs) and AI-2 (THMF) has been already discussed in this article. As for catecholamines, they apparently represent functional analogs of an additional autoinducer termed AI-3. Like catecholamines, AI-3 is an aromatic compound detected in GI tract-inhabiting bacteria, e. g., non-pathogenic and pathogenic *E. coli* strains, *Klebsiella pneumonia, Enterobacter cloacaceae, Shigella spp.*, and *Salmonella spp.* (Sircili et al., 2004; Walters et al., 2006). The effects of AI-3 result from its binding to two-component quorum-sensing systems whose receptors are termed QseC and QseE. AI-3 binding causes the phosphorylation of the response regulators QseB and QseF. Upon phosphorylation, they function as kinases that phosphorylate the activators of transcription of the genes responsible for flagellar motility (the *flhDC* genes) and virulence (the *LEE* genes) in the enterohemorrhagic *E. coli* strain O157:H7 (Clarke and Sperandio, 2005). Evidence has been presented that QseC is capable of binding catecholamines in addition to AI-3 per se (Clarke et al., 2006), and QseE is presumably characterized by a similar capacity. These receptors, therefore, are regarded as bacterial analogs of the catecholamine-binding receptors of eukaryotic cells including neurons, even though QseC and QseE differ from the eukaryotic G proteins structurally.

Despite this difference, the bacterial quorum-sensing receptor QseC is similar to eukaryotic α-adrenoreceptors in functional terms. Studies with *E. coli* O157:H7, *Salmonella enterica,* and *Yersinia enterocolitica* revealed that its interaction with norepinephrine, epinephrine, and AI-3 is specifically blocked by the α-adrenergic antagonists phentolamine, phenoxybenzamine, and prazosin, but not by the β-adrenergic antagonists propranolol and labetalol (Clarke et al., 2006; Freestone et al., 2007b). Dopamine also loses its stimulatory effect if added in combination with chlorpromazine that selectively blocks D_2 receptors. However, the dopamine effect is still observed if raclopride, a specific D_1 receptor antagonist, or haloperidol, a non-selective antagonist of both dopamine receptors in eukaryotic systems, are added (Freestone et al., 2007b). The fact that the effects of dopamine and norepinephrine are suppressed by different antagonists suggests that they are bound by different receptor sites of bacterial cells. This suggestion is consistent with the above data on the different effects of these two catecholamines on microbial growth and microcolony formation (Anuchin et al., 2008).

The AI-3- and catecholamine-perceiving systems of GI microflora presumably perform two different functions:

- They contribute to the cross-talk among various microorganisms in the GI tract, because AI-3 is an interspecies signal molecule. Moreover, bacteria synthesize norepinephrine and dopamine and release them into the culture fluid (Shishov et al., 2009), an issue to be discussed in the final section of this article. In particular, pathogenic *E. coli* strains can receive growth-, virulence-, and biofilm formation-stimulating signals from the commensal microflora including nonpathogenic *E. coli* strains.

• They are involved in the chemical dialog between the microflora and the host organism that specifically produces norepinephrine and epinephrine and releases them into the bloodstream (from which they penetrate into the intestinal lumen) in response to infection. We should re-emphasize that it is pathogenic microorganisms that are more responsive to norepinephrine than to dopamine. In contrast, nonpathogenic *E. coli* prefers dopamine to norepinephrine (Anuchin et al., 2008); it also responds to other neurotransmitters that are characteristic of local inflammation rather than of serious systemic infection (see below).

Adrenergic and dopamine receptor antagonists are of obvious medical interest. The data presented above "indicates that adrenergic antagonists can inhibit the AI-3/epinephrine/norepinephrine signaling cascade in EHEC and render it unable to induce its virulence genes (in response to these signals), suggesting that antagonists targeting this signaling cascade might constitute a novel class of antimicrobials," (Clarke et al., 2006).

2. Serotonin (5-Hydroxytryptamine), a derivative of the amino acid tryptophan, functions as a neurotransmitter in the nervous system of humans and various animal species. It combines its neurotransmitter function with the role of a local histohormone. As a neurotransmitter, serotonin is formed in the raphe nuclei of the brainstem and released into synaptic clefts by neuron axons in most major brain structures. Serotonin limits the spreading of excitation waves in the brain caused by stimulus perception. As a result, stimulus processing is normally compartmentalized in specialized loci within the brain. This compartmentalization is prevented by lysergic acid diethylamide (LSD), which disrupts the operation of serotonergic perceptive zones and, therefore, causes hallucinations. Serotonin can also "put the brain asleep," and the serotonin-releasing raphe nuclei belong to the sleep-inducing zones of the brain (Dubynin et al., 2003).

The effects of serotonin in microbial systems have not been as extensively studied as those of catecholamines. In contrast, serotonin does not stimulate the growth of such intestinal pathogens as the enterohemorrhagic strain (EHEC) of *E. coli* (M. Lyte, personal communication). Serotonin only slightly stimulates the growth of the pathogen *Aeromonas hydrophila* at a very high concentration (1 mM) (Kinney et al., 1999).

Datas has been obtained that shows micromolar serotonin concentrations stimulate cell proliferation and biomass accumulation in the non-pathogenic *E. coli* K-12 strain, although to a lesser extent than dopamine (Oleskin et al., 1998; Anuchin et al., 2008). In contrast to catecholamines, the serotonin effect is characterized by a bell-shaped, non-linear, concentration dependence curve; the maximum stimulation of *E. coli* growth was attained with ~1 µM serotonin. It stimulates the growth of the GI tract-inhabiting gram-positive bacterium *Enterococus faecalis* (Strakhovskaya et al., 1993). Serotonin also promotes the growth of microorganisms that do not belong to human/animal symbionts (or parasites). It accelerates the growth of the purple phototrophic bacterium *Rhodospirillum rubrum* (Oleskin et al., 1998), the soil myxobacterium *Polyangium sp.* (Oleskin et al., 1998), and the yeasts *Candida guillermondii* (Strakhovskaya et al., 1993) and *S. cerevisiae* (Malikina et al., 2010).

Serotonin exerts a stimulatory influence upon plants. For example, it accelerates the germination of radish seeds (Roschina, 1991). Presumably, serotonin produces its effects in plant systems because its chemical structure is similar to that of the plant hormone auxin, or 3-indolacetic acid (IAA). However, studies with *Pesudomonas putida* provided evidence that IAA is utilized by bacteria only as a carbon, nitrogen, and energy source (Leveau and Lindow, 2005).

The initial biofilm formation stage, cell aggregation and the development of matrix-embedded microcolonies, is stimulated by serotonin at a concentration of ~1 µM (Oleskin et al., 1998; Anuchin et al., 2008). Interestingly, higher serotonin concentrations (25–100 µM and above) cause cell dispersal and inhibit matrix synthesis in *E. coli* and *Polyangium sp.* In contrast to micromolar concentrations, submillimolar concentrations of serotonin do not significantly stimulate microbial growth and, with *E. coli*, may even slightly inhibit it (Oleskin et al., 1998).

Evidence has been presented that serotonin and the chemically similar compound melatonin (synthesized in the pineal gland from serotonin) efficiently suppress the development of the intracellular pathogens chlamydia (Rahman et al., 2005).

The mechanism of action of serotonin is not well understood. In an analogy to catecholamines, one can assume the existence of serotonin-binding bacterial cell receptors, which could bind serotonin-like compounds, apart from serotonin.

As far as serotonin-like compounds are concerned, **indole** (the "backbone" of the molecule of serotonin, an indolamine) seems to be of special importance because it and its derivatives are widespread in the world of microorganisms. In *E. coli,* indole suppresses biofilm formation and, in particular, the initial stage of this process in the GI tract – the attachment of bacteria to epithelial cells. Indole also decreases bacterial motility (Bansal et al., 2007; Lee et al., 2007b). It behaves like an antagonist of catecholamines that enhance biofilm formation, adherence, and cell motility. Mutation of the two *E. coli* genes, *yliH* and *yceP,* that have been shown to participate in indole synthesis, enhances biofilm formation, and this effect is reversed by extracellular indole (Domka et al., 2006). In contrast to *E. coli,* indole in *P. aeruginosa* and *P. fluorescens* stimulates biofilm formation (Lee et al., 2007b). Indole also influences many of the quorum-sensing phenotypes and virulence factor production in *P. aeruginosa* (T.K. Wood, personal communication, quoted from: Ryan, Dow, 2008). It induces multidrug exporter genes (*acrDE* and *cusB*) in *E. coli* K-12 (Hirukawa et al., 2005) and regulates virulence-related genes (the pathogenicity island on the chromosome) in pathogenic *E. coli* (Anyanful et al., 2005).

Studies with *E. coli* revealed that the indole derivatives 7-hydroxyindole and 5-hydroxyindole inhibit biofilm formation. However, another indole derivative, isatin (1-H-indole-2,3-dione), stimulated this process in pathogenic *E. coli* (EHEC) and did not effect the nonpathogenic *E. coli* (K-12). 2-Hydroxyindole produced no effect in all tested *E. coli* strains (Lee et al., 2007c).

The mechanism of action of indole is based upon the operation of a peculiar quorum-sensing system that is similar to N-acetylated homoserine lactone-dependent *luxI-luxR*-type systems briefly considered above. Such bacteria as *E. coli* or *Salmonella enterica* do not produce N-AHLs but respond to these signals when released by other bacteria. Apparently, N-AHLs are involved in interspecies communication proceedings, e. g. within the bacterial association of the GI tract. Foreign signals of the AI-1 type bind to protein SdiA, a homolog of protein LuxR in the classic *Vibrio fischeri* system. Protein SdiA also binds indole that behaves, therefore, as a functional analog of N-AHLs (reviewed, Ryan and Dow, 2008; Karatan and Watnick, 2009; Shpakov, 2009). In a recent work, protein enginnering has been used to reconfigure SdiA so as to make it either reduce or stimulate biofilm formation, depending on the mutants obtained (Lee et al., 2009).

It is an open question whether the effects of serotonin are due to the fact that it functions as an indole agonist (analog) and binds to an SdiA-type protein. Presumably, indole-binding bacterial proteins interact with indole derivatives formed by eukaryotes, such as indole-3-acetic acid, serotonin and melatonin, and these interactions influence the viability and pathogenicity of the bacteria involved. In addition, interspecies communication among bacteria within a biofilm can involve SdiA-type proteins and use N-AHLs and indole. While host-produced catecholamines stimulate the growth of GI microlora and biofilm formation, indole may convey the message, "This niche is occupied," and allow new bacterial cells to adhere only to vacant sites of the substratum where indole concentration is below a critical threshold value (Bansal et al., 2007).

3. Histamine, a derivative of the amino acid histidine. Histamine combines two functions. It operates as a neurotransmitter in a small zone of the hypothalamus and also functions as a histohormone involved in local inflammation (that also results in releasing serotonin). The data obtained by our team indicate that histamine efficiently stimulates cell proliferation and biomass accumulation in the nonpathogenic *E. coli* K-12 MC4100 strain. Like serotonin, histamine is characterized by a bell-shaped concentration dependence with a maximum effect at a concentration of 0.1 µM that stimulates *E. coli* growth twofold. Histamine promotes *E. coli* cell aggregation with microcolony formation (Anuchin et al., 2008), an early stage of a biofilm's life-cycle requiring a solid substratum (in our experiments, this substratum was the agar surface). Compared to serotonin, histamine loses its aggregation-promoting effect at higher concentrations (over 1 µM).

Micromolar histamine concentrations stimulate cell proliferation in *S. cerevisiae*, but, in contrast to the *E. coli* system, the stimulatory effect of histamine did not exceed that of serotonin and was significantly weaker than the effect of dopamine (Malikina et al., 2010).

The above data on the effects of biogenic amines on *E. coli* K-12 highlight the differences between the properties of the pathogenic and nonpathogenic (commensal) strains of *E. coli*. Norepinephrine was the most efficient growth-stimulating, cell adhesion- and biofilm formation-promoting agent with pathogenic strains such as EHEC. Since norepinephrine is produced during stress caused by infection, the response of EHEC and other pathogenic intestinal microorganisms is to be regarded as an evolutionary adaptation. It enables the pathogens to use a product of the host's protective response to accelerate their own development (Lyte, 1993, 2004; Freestone et al., 2007a).

Based on our data, the nonpathogenic symbiotrophic strain *E. coli* K-12 MC 4100 prefers a different neuromediator "landscape". Serotonin that is normally contained in the chromaffine granules of GI mucosa cells is not less efficient than norepinephrine in all tested systems (on the liquid and the solid medium). Dopamine, the minor component of the response to infection, stimulates proliferative activity and biomass accumulation to a greater extent than the major component norepinephrine. Histamine, a characteristic of local inflammation, was the most efficient growth-stimulating agent among the tested neurotransmitters. Presumably, the nonpathogenic strain *E. coli* K-12 MC 4100, in contrast to the pathogenic strain(s), is adapted, not to serious infection, but to light local inflammation. It is characterized by the synthesis and release of histamine, serotonin and, to a lesser extent, catecholamines that extrude into the intestinal lumen from nerve terminals damaged by inflammation. The local inflammation of the intestinal mucosa may be due to microtraumas caused by rough food.

4. Neuropeptides. Peptides in nervous system predominantly function as neuromodulators: they increase/decrease the efficiency of signal transmission across synapses whose operation depends on other neurotransmitters[4]. For instance, opiods (endorphins, enkephalins, and dinorphins) bind to specific neuron receptors and block impulse transmission along neuron axons, including those involved in pain perception. Opiods as pain-killers and euphoria-causing drugs represent an internal reward for an individual engaging in behavior resulting in opioid release. Opiods produced by the brain serve as positive reinforcement of altruistic acts: their production encourages law-abiding people even in situations in which obeying the law causes negative consequences for the individual involved (Gruter, 1991).

There are three opiod types that bind to μ-, δ-, and κ-opioid receptors. Zaborina et al. (2007) investigated the effects of these opiod types on the virulence of *P. aeruginosa* PAO1 that was estimated by monitoring pyocyanine synthesis. This research was carried out because preliminary studies demonstrated that the color of *P. aeruginosa* colonies became intensely green (indicative of pyocyanine production) after exposure of *P. aeruginosa* to filtered intestinal contents of stressed mice which contain opioids released from the richly innervated intestinal mucosa.

It was established in studies with different opiod types that only κ-opiod receptor agonists dinorphin and its synthetic analog U50,488 cause a considerable stimulation of pyocyanine production. The effect increases with an increase in κ-agonist concentration and results in an approximately 4-fold stimulation of pyocyanine production at a U50,488 concentration of 1 mM. The concentration dependence of the stimulatory effect of the μ-opiod

---

[4] Some neuropeptides serve as neurotransmitters themselves: they transmit impulses from neuron to neuron. For instance, substance P is responsible for transmitting pain-related impulses (for nociception).

receptor agonist morphine was bell-shaped, and the maximum stimulation level attained with 50 μM morphine was relatively low (40% of the control). However, at low population densities of the *P. aeruginosa* culture, the morphine effect increased, corresponding to a 6-fold stimulation of pyocyanine formation with 50 μM morphine. Finally, BW373U86, a synthetic δ-opiod receptor agonist, exerted an inhibitory effect on pyocyanine formation in *P. aeruginosa* (Zaborina et al., 2007).

The stimulatory effect of dinorphin and its analog U50,488 on *P. aeruginosa* virulence also resulted in enhancing the *P. aeruginosa*-caused suppression of the growth of useful GI microsymbionts such as the probiotics *Lactobacillus plantarum* and *L. rhamnosum*, and of the reproduction of the flatworm *Caenorhabditis elegans*.

While catecholamines behave as analogs of AI-3, dinorphin and its synthetic analog, in a similar fashion, perform the functions of the quorum-sensing autoinducer quinolone. The quinolone-dependent quorum-sensing system in *P. aeruginosa* is activated by the *lasI-lasR* system of this bacterium. Both quorum-sensing systems are involved in virulence factor synthesis and biofilm formation. It was demonstrated that the effects of dinorphin and U50,488 require the operation of these quorum-sensing systems. Mutations disrupting these systems prevent their effects. Moreover, the addition of dinorphin results in enhancing gene *pqsABCDE* expression, which stimulates the synthesis of 3,4-dihydroxy-2-heptylquinoline, 4-hydroxy-2-quinoline, and 2-heptyl-4-hydroxyquinoline-N-oxide, the autoinducers of the quinolone-dependent quorum-sensing system (Zaborina et al., 2007).

Opiods accumulate during stress in the human/animal organism, including the GI tract. It is to be expected that their activating influence on the quinolone-dependent quorum-sensing system increases the virulence of *P. aeruginosa*, resulting in the suppression of the growth of non-pathogenic enterobacteria and the colonization of the GI tract by *P. aeruginosa* (Shpakov, 2009).

Taken together, the data on various neurotransmitters and their analogs suggest an important mechanism that links stress and the development of bacterial infection often accompanied by biofilm formation in the human/animal organism. In summary, vertebrate neurotransmitters whose synthesis and release are stimulated by stress factors can behave in an autoinducer-like fashion if they contact bacterial cells (Shpakov, 2009, p.170).

5. Nitric oxide (NO), a highly reactive gaseous free-radical substance, belongs to the "gasotransmitters," i.e. gases which function as neurotransmitters and/or hormones. As a hormone, NO is responsible for the dilation of blood vessels. As a neurotransmitter, NO is involved in the operation of the brain zones that function during grooming behavior in animals and petting in humans. The positive emotional state characteristic of such situations is partly due to the effects of NO. In mice, a mutation in the *Nos1* gene that encodes the NO synthase making NO from arginine manifests itself in increased aggressivity and frequent attempts to mount sexually inactive females (Tecott and Barondes, 1996). In animal and other eukaryotic cells, NO is involved in cell differentiation, apoptosis (programmed cell death), and cell proliferation.

NO predominantly performs regulatory functions at low concentrations, whereas its high concentrations are toxic both for host cells and microbial symbionts or parasites. This cytotoxic effect of nitric oxide is used by the macroorganism in terms of the immune response to foreign or tumor cells. Immune cells from the bloodstream (macrophages) release NO, along with reactive oxygen species, and these "weapons" kill potentially dangerous cells. Cytotoxic NO concentrations are also produced by other cell types, including hepatocytes and endothelial cells. They use nitric oxide to destroy protozoans and helminths (James, 1995).

A serious infection results in NO production by the human/animal organism. NO produces a toxic effect on the whole organism, causing a life-threatening septic shock. The organism may kill itself before it is destroyed by the pathogen. According to V.P. Skulachev (1999), this prevents the pathogen from spreading in the population at the expense of the lives of some individuals in it.

Microorganisms respond to high (micromolar) NO concentrations using the stress response mechanism. This may account for the stimulatory effect of high NO concentrations on biofilm formation in *P. aeruginosa*[5] (Barraud et al., 2006), because biofilm formation, as mentioned above, partly results from the response of a microbial population to a stress factor.

However, low (pico- to nanomolar) NO concentrations are normally generated by various microbial species. NO formation is one of the stages of the denitrification process that proceeds as follows: $NO_3^- \rightarrow NO \rightarrow N_2O$

---

[5] A fivefold biofilm formation increase was achieved in these experiments with 75 mM of sodium nitroprusside, a nitric oxide donor (Barraud et al., 2006). The actual NO concentration was 1000 times lower (Barraud et al., 2009a), i.e. 75 µM.

(Zumft, 1993). Like eukaryotes, prokaryotes use NO (at low concentrations) as a regulatory agent.

In contrast to high NO concentrations, low NO concentrations inhibit biofilm formation and cause biofilm dispersal, which . are involved in the last stage of a biofilm's life-cycle (see the beginning of the article). In studies on the effects of NO in microbial systems, NO donors were added to the biofilms of pathogenic or opportunistic microorganisms inhabiting various niches of the human organism, invluding *P. aeruginosa* (Barraud et al., 2006), *Serratia marcescens, Vibrio cholerae, E. coli* (the pathogenic strain BW20767), *Staphylococcus epidermidis,* the yeast *Candida albicans* (Barraud et al., 2009b), and the food-spoiling bacterium *Bacillus licheniformis* (Barraud et al., 2009b). In the single-species biofilms of the tested microorganisms and in multispecies biofilms from water distribution and treatment systems, nitric oxide caused, on average, a 63% reduction in biofilm size.

Interestingly, the difference between the high- and low-concentration effects is also characteristic, as mentioned above, of the biogenic amines serotonin and histamine.

Low NO concentrations enhance the capacity of antimicrobial compounds, e. g., the antibiotic tobramycin, hydrogen peroxide, and the detergent sodium dodecylsulfate, to remove microbial biofilms from water distribution and treatment systems (Barraud et al., 2006, 2009b). Even "the efficacy of conventional chlorine treatments at removing multi-species biofilms from water systems was increased by 20-fold in biofilms treated with NO compared with untreated biofilms. These data suggest that combined treatments with NO may allow for novel and improved strategies to control biofilms and have widespread applications in many environmental, industrial and clinical settings," (Barraud et al., 2009b, p.370). This data can be explained in two ways: (i) The access of antimicrobial agents to microbial cells within biofilms is facilitated by biofilm dispersal in the presence of NO and (ii) Biofilms are regarded in related literature as analogs of stationary-phase cultures, and nitric oxide activates the genetic systems that induce the transition to the active growth stage, which increases the sensitivity of bacterial cells to toxic agents (Barraud et al., 2006).

The data on the influence of exogenous NO on microbial biofilms are not sufficient for proving the fact that NO produces a regulatory effect on them. The role of endogenous NO is to be investigated as well. Of special interest in this context are studies with mutant bacterial strains. It was established that the biofilm of the NO synthase-deficient *ΔnirS* mutant of *P. aeruginosa* did not disperse after 6 days of cultivation, in contrast to the wild-type strain. The biofilm of the *ΔnorCB* mutant lacking the nitric oxide-removing enzyme NO reductase dispersed to a greater extent than the wild-type biofilm, so that numerous hollow voids were formed and cell death was enhanced (Barraud et al., 2006). Low, nontoxic levels of NO in *P. aeruginosa* biofilms upregulate the *pilA* gene involved in swarming and twitching motility and downregulate genes encoding adhesins (the *cupB* and *cupC* genes) and virulence factors, such as the pyoverdine-encoding *pvd* gene (Barraud et al., 2009a).

It was revealed that the regulatory influence of NO on biofilms is due to its effect on the c-di-GMP level (Barraud et al., 2009a). NO decreases the intracellular c-di-GMP pool in *P. auruginosa* by activating the c-di-GMP-degrading enzyme PDE. It was demonstrated that inhibiting thie enzyme with GTP reduces the biofilm dispersal-promoting effect of NO. The NO influence on c-di-GMP is mediated by intracellular protein BdlA (a chemotactic transducer): in the *bdlA* mutant lacking the gene encoding this protein, the intracellular c-di-GMP concentration did not decrease in the presence of an NO donor. Taken together, the data obtained testify to a similarity between the regulatory effects of NO in microbial biofilms and eukaryotic multicellular systems with respect to cell differentiation, proliferation, and programmed death. This similarity is highlighted by the fact that a large number of factors involved in NO-dependent regulatory processes are evolutionary conserved proteins.

The data concerning the regulatory effects of NO on biofilm dispersal in symbiotic and/or parasitic microorganisms give grounds for the suggestion that it is involved in the chemical communication between them and the host organism that contains various types of cells capable of producing NO (James, 1995). The fact that NO is produced by a wide variety of microbial species enables it to function as an interspecies signal molecule within the microbial association inhabiting the GI tract and other niches in the human/animal organism. The implications of other "gasotransmitters" such as CO and H_2S for microbial symbiotic biofilms are to be addressed in further studies in this field.

The limited volume of this work prevents a detailed discussion concerning the microbiological implications of other neurotransmitter groups such as amino acids. Suffice to mention that the macro- and microstructure of *E. coli* colonies and, presumably, biofilms are formed under the influence of aspartic acid (Budrene and Berg, 1991, 2002) as an attractant. Complex structural patterns including concentric circles and hexagonal lattices result from the superposition of two concentration gradients of aspartic acid: (i) formed in the colony center and (ii) produced by the cells on the colony periphery. Bacteria form concertedly moving clusters that generate these complex patterns in the presence of aspartic acid (Mittal et al., 2003). Aspartic acid is an evolutionarily conserved signal molecule representing one of the major neurotransmitters in mammals and other animals.

Chapter VI

Neurotransmitter Biosynthesis and Release by Microorganisms: Implications for Symbiotic Biofilms

The preceding section, apart from the influence of exogenous neuromediators, actually dealt, with their production by microbial cell , at least in the example of nitric oxide. Do microorganisms synthesize other neurotransmitters, e. g., biogenic amines? Or are they only capable of producing chemically similar analogs of neurotransmitters, exemplified by the catecholamine analog AI-3 (Freestone et al., 2007a and other references given above) and the serotonin analog indole (Lee et al, 2007b, c etc.)?

The data presented up to now concerning the occurrence of neurotransmitter amines in microorganisms are far from being sufficient. Most of them concern unicellular eukaryotes. Detailed information on their contents in some groups of eukaryotic microorganisms, e.g., in protozoans, is available in related literature (reviewed, Buznikov, 1987, 2007). The data on biogenic amines in prokaryotes are fragmentary. Serotonin was detected in *Enterococcus faecalis* (Strakhovskaya et al., 1993). It is also synthesized by some of the intestinal bacteria of parasitic nematodes (Hsu et al., 1986), including *Staphylococcus aureus* that produced the maximum serotonin amount. Evidence was presented in literature stating that human GI microflora synthesizes and releases serotonin, histamine, and neurotransmitter amino acids such γ-aminobutyric acid (GABA), glutamic acid, β-alanine, etc. (Babin et al., 1994; Lazdin et al., 1999). Microbial

GABA functions as an inhibitory neurotransmitter involved in blocking the transmission of impulses in a number of parts of the nervous system. Severe stress or the effects of antibiotics change the composition of the microflora, resulting in decreased GABA synthesis and an increased pain sensitivity of the colon (Babin et al., 1994). This causes irritated bowel syndrome and probably plays a role in the development of inflammatory bowel diseases such as ulcerous colitis and Crohn's disease.

The data obtained by our team using high performance liquid chromatography (HPLC) with electrodetection will be briefly discussed below. We demonstrated that the biomass of the gram-positive bacteria *Bacillus cereus* and *Staphylococcus aureus* contains serotonin at micromolar concentrations (Tsavkelova et al., 2000). For comparison, it should be noted that serotonin occurs at similar concentrations in the bloodstream of mammals (Kruk and Pycock, 1990).

We also established that the catecholamines norepinephrine and dopamine are actually widespread in the microbial world (Tsavkelova et al., 2000), and their concentrations in the biomass of some of the tested microorganisms exceeded that in the mammal bloodstream. Norepinephrine was detected in the yeast *S. cerevisiae* and the penicillin-producing fungus *Penicillium chrysogenum*.

In the example of the matrix-rich bacterium *Bacillus subtilis*, (the M variant) we demonstrated that neurotransmitters (norepinephrine and dopamine) concentrate in the matrix fraction. This fact supports the idea that these amines function as intercellular signals, because the hydrophilic biopolymer components of the matrix promote the diffusion of low molecular weight signal molecules within the colony (biofilm). If this idea is valid, neurotransmitters should be regarded as short-range signal molecules not only in animals where they transmit impulses from neuron to neuron across the narrow synaptic cleft, but also in microorganisms, because the matrix retains the neurotransmitters within the boundaries of the colony/biofilm whose cells have produced them.

A majority of the tested microorganisms also contain the products of oxidative deamination of biogenic amines such as 5-hydroxyindolylacetic acid (5-HIAA) and dihydrophenylacetic acid (DHPAA). However, the data presented in the work by Tsavkelova et al. (2000) were obtained only with cells harvested during the late exponential phase of the tested microbial cultures.

In a later work (Shishov et al, 2009), we investigated the dynamics of the synthesis of the neurotransmitter amines serotonin, norepinephrine, and dopamine during the growth of an *E. coli* K-12 culture on the synthetic medium M-9 and, for comparison, on the rich Luria-Bertani (LB) medium. It was revealed that maximum (micromolar) concentrations of these compounds are contained in *E. coli* cells[6] during the intial growth phases of the culture, i.e. during the lag phase or, in the case of dopamine produced on the LB medium, during the early exponential phase. Norepinephrine is the dominant neurotransmitter. The intracellular contents of the tested compounds decreased after the transition to the late growth phases. Importantly, these neuromediators were also detected in *E. coli* cells grown on the synthetic M-9 medium that contains no neuromediators. Hence, the neuromediators were synthesized inside the cells, and not taken up from the medium.

The data obtained should be considered in conjunction with the earlier established fact (Oleskin et al., 1998; Oleskin and Kirovskaya, 2006; Anuchin et al., 2008) that exogenous neurotransmitters are maximally efficient in stimulating cell proliferation during the initial growth phases of the culture. Presumably, endogenous neurotransmitter amines are present in bacterial cells at maximum concentrations during the growth phases in which they exert their strongest regulatory influence on the bacterial culture involved. Accordingly, the amines might function as "molecular switches" that "turn on" active growth and cell division during the initial growth phases.

E. coli biomass contained dihydroxyphenylalanine (DOPA) and 5-hydroxytryptophan (5-HTP) that serve as the precursors of catecholamines and serotonin, respectively, in animal cells. We also detected the products of their oxidative deamination, including 5-HIAA, DHPAA, and homovanilic acid. Taking account of these data, we suggested that the biosynthesis and degradation of neurotransmitter amines in bacterial cells apparently involve animal cell-specific enzymes or their homologs. Presumably, these processes are carried out as follows:

Tryptophan → 5-HTP → Serotonin → 5-HIAA
Tyrosin → DOPA → Dopamine → Norepinephrine[7]
↘ DHPAA → Homovanilic acid

---

[6] To determine intracellular concentrations of neurotransmitters, we disintegrated microbial cells by sonication.
[7] The scheme does not include the products of norepinephrine deamination, because they were not detected in our system

It seems likely, therefore, that the patterns of biosynthesis and degradation of neurotransmitter amines are common for pro- and eukaryotes. Presumably, bacteria form neurotransmitter-synthesizing enzymes such as hydroxylases and aromatic amino acid decarboxylases as well as enzymes involved in their degradation, including monoaminooxidases (MAOs).

The following is the data on the dynamics of the release of neurotransmitters, their precursors and oxidation products into the culture fluid during the growth of *E. coli* cultures. The culture fluid supernatant (CSF) of *E. coli* grown on the synthetic M-9 medium contains nanomolar serotonin, dopamine, and norepinephrine concentrations during the late growth phases (Shishov et al., 2009). These concentrations are sufficient to enable the neurotransmitters to bind to respective receptors in the GI tract of the human/animal host of *E. coli*. Importantly, the cell density during the later exponential phase (characterized by relatively high neurotransmitter concentrations in the CSF) was about 5×10^8 cells/ml, which is comparable to the cell density occurring in the human large intestine. It may contain over 10^8 colony-forming units of *E. coli* per ml (Shenderov, 2001). This data can apparently testify to possible physiological effects of the amines released by the intestinal symbiont *E. coli*. However, the above data was only obtained with suspension cultures that are comparable with the planktonic cells of the intestinal lumen. Presently, research is in progress on the neurotransmitter synthesis and release by the biofilms of *E. coli* and other symbiotic microorganisms, which should simulate the biofilms formed on the mucosa of the GI tract.

The CSF of *E. coli* lacked the serotonin precursor 5-HTP. However, of particular interest seems to be the fact that not only the biomass, but also the culture fluid contains over 1 µM DOPA, the catecholamine precursor, at the late stages of the growth of the *E. coli* culture.

Presumably, DOPA can function as a long-range signal by diffusing from one cell aggregate/microcolony to another. Its conversion to dopamine that stimulates bacterial cell proliferation (Anuchin et al., 2008) may be performed within a cell upon taking up DOPA molecules. The phenomenon that the lag phase is reduced by the CSF of another culture which has reached a later growth phase was first described about one hundred years ago (Rahn, 1904; Penfold, 1914). This phenomenon can be accounted for in terms of the postulated mechanism of action of DOPA, along with other autostimulators. Specifically, the DOPA released by an exponential-phase culture may be recognized and taken up by sensitive lag-phase cells.

The release of large DOPA amounts into the medium by the symbiont *E. coli* raises the question whether it can exert a considerable influence on the whole human organism including its nervous system. DOPA crosses the gut-blood and blood-brain barriers. In the brain, DOPA is converted to dopamine and thereupon to norepinephrine. It should be re-emphasized that dopamine and norepinephrine regulate important brain processes. They are involved in locomotion, affection, sociable and dominant behavior, as well as aggression. Hence the release of micromolar amounts of DOPA by *E. coli* cells enables it to exert an important influence on human psyche and social behavior.

We have recently conducted studies with the gram-positive bacterium *Bacillus cereus* using HPLC. Like *E. coli* K-12, it contained maximum (micromolar) norepinephrine and dopamine concentrations during the lag phase. However, DOPA was present at high concentrations (up to 10 μmoles/kg of biomass) during the late exponential phase. Only nanomolar serotonin concentrations were detected by us in *B. cereus* cells. Apart from the biomass , the culture fluid supernatant contained micromolar norepinephrine and DOPA amounts during the late growth phases (Shishov, V.I. & Oleskin, A.V., unpublished). *B. cereus* grows on food items and can cause food poisoning. These findings suggest a potential contribution of catecholamines (that stimulate the growth of pathogenic enterobacteria) and the blood-brain barrier-crossing DOPA to the effects of *B. cereus* on the human organism.

In general, microorganism-produced neurotransmitters and their precursors and products, along with other biologically active substances, can produce local and systemic effects in the organism.

Apart from neurochemicals and related substances, special emphasis should be placed upon amino acids, proteins (including enzymes), vitamins including those of the B group, carbon acids, cyclic nucleotides (cAMP and cGMP), and short-chain fatty acids. These chemical agents produced by symbiotic microflora and its biofilms influence a large number of important processes carried out by the GI tract and the human organism as a whole, which has important clinical, psychological, and biopolitical implications.

Presently, studies are in progress in our lab that are aimed at investigating the dynamics of biosynthesis and release of neurotransmitter amines in the yeast *S. cerevisia*. The results already obtained (Malikina et al., 2010) indicate that norepinephrine, dopamine, and serotonin, as well as their oxidation products homovanilic acid, DHPAA, and 5-HIAA and the catecholamine precursor DOPA are contained in yeast cells at micromolar or submicromolar concentrations. Nevertheless, they are not released into the culture fluid supernatant. On the neurotransmitter-containing peptone–yeast extract–maltose medium, the CSF concentrations of the tested compounds tend to decrease during the growth of the *S. cerevisiae* culture, suggesting that yeast cells actively take them up from the medium. As for the synthetic, neurotransmitter-lacking sucrose–ammonium medium, the intracellular accumulation of neurotransmitters does not result in their release into the culture fluid. The data obtained appear to militate against the idea that neurotransmitter amines operate as autoregulators in *S. cerevisiae* populations. However, they may be involved in regulating the growth of the yeast population by other components of an ecosystem. We should re-emphasize the significant growth-stimulating effects produced on yeast cells by the neurotransmitters, particularly by dopamine (Malikina et al., 2010).

In nature, yeast forms biofilms on the skin of plums, grapes, and other kinds of fruit. It is to be expected that *S. cerevisiae* cells exchange signal molecules with other pro- and eukaryotic microorganisms within the multispecies biofilms, as well as with the plant host. Presumably, neurotransmitter amines are unidirectional signals received by yeast and emitted by other ecosystem components. These components may include bacteria that release neurotransmitters and their precursor DOPA into the medium, according to the above data.

Like human organism-produced catecholamines that are recognized by enterobacteria as analogs (agonists) of their own quorum-sensing factor AI-3, neurotransmitter amines and DOPA can serve as functional analogs of yeast-synthesized factors that resemble them in structural terms. Such yeast-synthesized regulatory agents include aromatic alcohols. For example, the alcohols triptophol (a derivative of tryptophan, a precursor of 5-HTP) and phenylethanol (a derivative of phenylalanine, which converts into DOPA) function as autoregulators in *S. cerevisiae*. They are formed by *S. cerevisiae* cells under nitrogen limitation and trigger their transition from solitary cells to branched pseudomycelium (Chen and Fink, 2006). Another DOPA-related compound, the alcohol tyrozol is an autoregulator formed by the yeast *Candida albicans,* a potentially pathogenic inhabitant of the human organism (Chen et al., 2004).

Chapter VII

Conclusion

The functional roles of neurotransmitters in respect to microorganisms and their biofilms as well as their probable involvement in the communication between symbiotic microflora and the host organism form part of the agenda of the area of research termed **microbial endocrinology** (Lyte, 1993, 2004). This subfield of microbiology concentrates on the functions of interorganismic signal molecules of animals and humans in unicellular organisms, with special attention to their functional differentiation, social behavior, and communication. Apart from neurotransmitters, microbial endocrimology focuses on substances that function as hormones in animals and are present in microorganisms. For example, the pancreas hormone insulin was detected in a large number of microorganisms including the fungus *Neurospora crassa*. In this fungus, insulin is involved in regulating carbon metabolism. *N. crassa* contains a gene that is homologous to the insulin gene of mammals (Lenard, 1992). Insulin was found to decrease adenylate cyclase activity in *E. coli* M-17, whereas the thyroid hormone thyroxin was without effect and epinephrine increased adenylate cyclase activity, although only in a stationary-phase culture (Korobov et al., 1976).

Importantly, "the presence of such hormones in micro-organisms is believed to represent a form of intercellular communication and as such may constitute a type of primitive nervous system," (Lyte, 1993, p.343).

Besides the existence of similar or identical mediators in neuron networks and microbial systems including biofilms, special emphasis should be placed on the data concerning the presence of neurotransmitter receptor homologs in microorganisms. Apart from the operation of catechol-binding QseC and QseE receptors in a number of microorganisms, it should be mentioned that the purple phototrophic bacterium *Rhodobacter sphaeroides* contains CrtK, a homolog of the benzadipine receptor binding γ-aminobutyric acid (Baker and Fanestil, 1991). It is generally accepted that the mitochondria of eukaryotic cells have descended from prokaryotic ancestors belonging to the subgroup represented by *R. sphaeroides*. Therefore, research on bacterial receptors binding neurotransmitters is of considerable importance in terms of brain neurochemistry, taking into account the data on the role of brain neuron mitochondria in binding neurotransmitters. Neuron mitochondria contain, for example, glutamate receptors of the NMDA type (Montal, 1998). If glutamate is present at high concentrations, its binding to these mitochondrial receptors results in the intrusion of a large amount of Ca^{2+} ions into the mitochondria, the collapse of their membrane potential, a decrease in intracellular ATP concentration, and, finally, apoptosis. Brain neuron apoptosis caused by high concentrations of glutamate and other neurotransmitters probably occurs in patients suffering from neurodegenerative diseases such as Parkinson's, Huntington's, and Altzheimer's disease.

In fact, microbial biofilms resemble neuron networks in the animal/human nervous system in several respects. Apart from the existence of similar/homologous/identical signal molecules and receptors, networks of microbial cells within biofilms bear structural similarity to neuron networks and might be capable of collective information processing and decision-making as well.

As for the biofilms that are formed by human symbiotic micro organisms, their "thinking networks" (albeit primitive) are highly responsive to the health state, stress level, and even mood of a human individual. Since the state of an individual is influenced by his interactions with others in human society, the microbial symbionts apparently should respond to the whole social atmosphere, including the political situation. Therefore, their activities and ongoing communication with the human host, which involves various biologically active substances including neurotransmitters, is of considerable interest in biopolitical terms.

References

Anuchin, A. M., Chuvelev, D. I., Kirovskaya, T. A., & Oleskin, A.V. (2008). Effects of monoamine neuromediators on the growth-related variables of *Escherichia coli* K-12. *Mikrobiologiya, 77,* 674-680.

Anyanful, A., Dolan-Livengood, J. M., Lewis, T., Sheth, S., Dezalia, M. N., Sherman, M. A., Kalman, L. V., Benian, G. M., & Kalman, D. (2005). Paralysis and killing of *Caenorhabditis elegans* by enteropathogenic *Escherichia coli* requires the bacterial tryptophanase gene. *Mol. Microbiol., 57,* 988-1007.

Babin, V. N., Domaradsky, I. V., Dubinin, A. V., & Kondrakova, O. A. (1994). Biochemical and molecular aspects of the symbiosis of the human being and his microflora. *Ross. Khim. Zhurn., 38,* 66-78.

Bai, A. P. & Ouyang Q. (2006). Probiotics and inflammatory bowel diseases. *Postgraduate Med. J.* 82, 376-382.

Baker, M. E. & Fanestil, D. D. (1991). Mammalian peripheral-type benzodiazepine receptor is homologous to CrtK protein of *Rhodobacter capsulatus*, a photosynthetic bacterium. *Cell, 65,* 721-722.

Bansal, T., Englert D., Lee, J, Hegde, M., Wood, T. K., & Jayaraman, A. (2007). Differential effects of epinephrine, norepinephrine, and indole on *Escherichia coli* O157:H7 chemotaxis, colonization, and gene expression. *Infect. Immun., 75,* 4597-4607.

Barraud, N., Hassett, D. J., Hwang S.-H., Rice, S. A., Kjelleberg, S., & Webb, J. S. (2006). Involvement of nitric oxide in biofilm dispersal of *Pseudomonas aeruginosa. J. Bacteriol., 188,* 7344-7353.

Barraud, N., Schleheck, D., Klebensberger, J., Webb, J. S., Hassett, D. J. Rice, S. A., & Kjelleberg, S. (2009a). Nitric oxide signaling in *Pseudomonas aeruginosa* biofilms mediates phosphodiesterase activity, decreased cyclic di-GMP levels, and enhanced dispersal. *J. Bacteriol,.* 191, 7333-7342.

Barraud, N., Storey, M. V., Moore, Z. P., Webb, J. S., Rice, S. A., & Kjelleberg, S. (2009b). Nitric oxide-mediated dispersal in single- and multi-species biofilms of clinically and industrially relevant microorganisms. *Mol. Microbiol., 2,* 370-378.

Berg, R. D. (1996). The indigenous bacterial microflora. *Trends Microbiol. 4,* 430-435.

Berridge, K. C. & Robinson, T. E. (1998). What is the role of dopamine in reward: hedonic impact, reward learning or inentive salience? *Brain Res. Rev., 28,* 309-369.

Budrene, E. O. & Berg, H. C. (1991). Complex patterns formed by motile cells of *Escherichia coli. Nature, 349,* 630-633

Budrene, E. O. & Berg, H. (2002). Dynamics of formation of symmetrical patterns by chemotactic bacteria. *Nature, 376,* 49-53.

Burton, C. L., Chhabra, S. R., Swift, S., Baldwin, T. J., Withers, H., Hill, S. J., & Williams P. (2002). The growth response of *Escherichia coli* to neurotransmitters and related catecholamine drugs requires a functional enterobactin biosynthesis and uptake system. *Infect. Immunol., 70.* 5913-5923.

Buznikov, G. A. (1987). Neirotransmittery v Embriogeneze (Neurotransmitters in Ontogeny). Moscow: Nauka.

Buznikov, G. A. (2007). Preneural transmitters as regulators of embryogenesis. Present-day state-of-the-art. *Ontogenez, 38,* 262-270.

Camilli A. & Bassler, B. L. (2006). Bacterial small-molecule signaling pathways. *Science, 311,* 1113-1116.

Chen, C., Brown, D. R., Xie, Y., Green, B. T., & Lyte, M. (2003). Catecholamines modulate *Escherichia coli* O157:H7 adherence to murine cecal mucosa. *Shock, 20,* 183-188.

Chen, H., Fujita, M., Feng, Q., Clardy, J., & Fink, G. R. (2004). Tyrosol is a quorum-sensing molecule in *Candida albicans. Proc. Natl. Acad. Sci. USA, 101,* 5048-5052.

Chen H. & Fink, G. P. (2006). Feedback control of morphogenesis in fungi by aromatic alcohols. *Genes Dev., 20,* 1150-1161.

Clarke, M.B. & Sperandio, V. (2005). Transcriptional autoregulation by quorum sensing Escherichia coli regulators B and C (QseBC) in enterohaemorrhagic E. coli (EHEC). *Mol. Microbiol., 58,* 441-455.

Clark, M. B., Hughes, D. T., Zhu, C., Boedeker, E. C., & Sperandio, V. (2006). The QseC sensor kinase: A bacterial adrenergic receptor. *Proc. Natl. Acad. Sci. USA, 103,* 10420-10425.

References

Davies, D. G., Parsek, M. R., Pearson, J. P., Iglewski, B. H., Costerton, J. W., & Greenberg, E. P. (1998). The involvement of cell-to-cell signaling in the development of a bacterial biofilm. *Science, 280*, 295-298.

Domka, J., Lee, J., & Wood, T. K. (2006). YliH (BssR) and YceP (BssS) regulate Escherichia coli K-12 biofilm formation by influencing cell signaling. *Appl. Environ Microbiol., 72*, 2449-2459.

Duan, K. M., Dammel, C., Stein, J., Rabin, H., & Surette, M. G. (2003). Modulation of *Pseudomonas aeruginosa* gene expression by host microflora through interspecies communication. *Mol Microbiol., 50*, 1477-1491.

Dubynin, V. A., Kamensky, A. A., Sapin, M. R., & Sivoglazov, V. N. (2003). *Regulyatornye sistemy organizma cheloveka (Regulatory Systems of the Human Organism)*. Moscow: Drofa.

Duda, V. I., Vypov, M. G., Sorokin, V. V., Mityushina, L. L., & Lebedinsky, A. V. (1995). Formation of hemoprotein-containing extracellular structures by bacteria. *Microbiologia, 64*, 69-73.

Duda, V. I., Il'chenko, A. P., Dmitriev, V. V., Shorokhova, A. P., & Suzina, N. E. (1998). Isolation and characterization of a hemopflavoprotein from the gram-negative bacterium *Alcaligenes* sp., strain d_2. *Mikrobiologiya, 67*, 12-18.

Freestone, P. P, Haigh, R. D., Williams P. H., & Lyte M. (1999). Stimulation of bacterial growth by heat-stable, norepinephrine-induced autoinducers. *FEMS Microbiol. Lett., 172*, 53-60.

Freestone, P. P, Haigh, R. D., & Lyte M. (2007a). Specificity of catecholamine-induced growth in *Escherichia coli* O157:H7, *Salmonella enterica* and *Yersinia enterocolitica*, *FEMS Microbiol. Lett. 269*, 221-228.

Freestone, P. P, Haigh, R. D., & Lyte M. (2007b). Blockade of catecholamine-induced growth by adrenergic and dopaminergic receptor agonists in *Escherichia coli* O157:H7, *Salmonella enteric*, and *Yersinia enterocolitica*. *BMC Microbiol., 7*, 8.

Freestone, P. P, & Lyte M. (2008). Microbial endocrinology: Experimental design issues in the study of interkingdom signaling in infectious disease. *Adv. Appl. Microbiol., 64*, 75-108.

Fuqua, W. C., Winans, S. C., & Greenberg, E. P. (1994). Quorum sensing in bacteria: the LuxR-LuxI family of cell density-responsive transcriptional regulators *J. Bacteriol., 176*, 269-275.

Gerstel, U. & Romling, U. (2003). The *csgD* promoter, a control unit for biofilm formation in *Salmonella typhimurium*. *Res. Microbiol., 154*, 659-667.

Gonzalez Barrios, A. F., Zuo, R., Hashimoto, Y., Yang, L., Bentley, W. E., & Wood, T. K. (2006). Autoinducer 2 controls biofilm formation in *Escherichia coli* through a novel motility quorum-sensing regulator (MqsR, B3022). *J. Bacteriol., 188*, 305-316.

Green, B. T., Lyte, M, Chen, C, Xie, Y., Casey, M. A., Kulkarni-Narla, A., Vulchanova, L., & Brown, D. R. (2004). Adrenergic modulation of *Escherichia coli* O157:H7 adherence to the colonic mucosa. *Am. J. Physiol. Gastrointest. Liver Physiol., 287*, G1238-1246.

Gruter, M. (1991). Law and the mind. Biological origins of social behavior. Newbury Park: L & New Delhi.

Hall-Stoodley, L., Costerton, J. W., & Stoodley, P. (2004). Bacterial biofilms: from the natural environment to infectious diseases. *Nat. Rev. Microbiol., 2*, 95-108.

Hawrelak, J. A.. The causes of intestinal dysbiosis: a review. 2009 2 7. Available from: http://findarticles.com/p/articles/mi m0FDN/is 2 9/ai n6112781/print?tag=artBody;col1

Hirukawa, H., Inazumi, Y., Masaki, T., Hirata, T., & Yamaguchi A. (2005). Indole induces the expression of multidrug exporter gene in *Escherichia coli. Mol. Microbol., 55*, 1113-1126.

Hoffman, L. R., D'Argenio, D. A., MacCoss, M. J., Zheng, Z., Jones, R. A., & Miller, S. I. (2005). Aminogycoside antibiotics induce bacterial biofilm formation. *Nature, 436*, 1171-1175.

Hsu, S. C., Johansson, K. R., & Donahne, M. J. (1986). The bacterial flora of the intestine of *Ascaris suum* and 5-hydroxytryptamine production. *J. Parasitol., 72*, 545-549.

Itoh, Y., Wang X., Hinnebusch B. J., Preston III, J. F., & Romeo, T. (2005). Depolymerization of beta-1,6-N-acetyl-D-glucosamine disrupts the integrity of diverse bacterial biofilms. *J. Bacteriol., 187*, 382-387.

James, S. L. (1995). Role of nitric oxide in parasitic infections. *Microbiol. Rev., 59*, 533-547.

Jefferson, K. K. (2004). What drives bacteria to produce a biofilm? *FEMS Microbiol. Lett., 236*, 163-173.

Kaper, J.B. & Sperandio, V. (2005). Bacterial cell-to-cell signaling in the gastrointestinal tract. *Infect. Immmun., 73*, 3197-3209.

Karatan, E. & Watnick, P. (2009). Signals, regulatory networks, and materials that build and break bacterial biofilms. *Microbiol. Mol. Biol. Rev., 73*, 310-347.

Kinney, K. S., Austin, C. E., Morton, D. S., & Sonnenfeld, G. (1999). Catecholamine enhancement of *Aeromonas hydrophila* growth. *Microbial Pathogenesis., 25*, 85-91.

Korobov, V. P., Kalyuzhnyi, V. M., & Dedyukina, M. M. (1976). Adenylate cyclase of *E. coli*: Hormonal activation and inhibition. In: *Conference on The Molecular Mechanisms of Hormone Action.* (p.30). Moscow: USSR Academy of Medical Sciences.

Kruk, Z. L. & Pycock, C. J. (1990). *Neurotransmitters and Drugs.* L., N.Y., Tokyo: Chapman & Hall.

Lazdin, O. A., Chervinets, V. M., & Tabakov, T. D. (1999). *Mikrobiotsenoz Kishechnika i Ego Korrektsiaya (Intestinal Biocenosis and Its Amelioration).* Tver: State Medical Academy of Tver.

Lee, V. T., Matewish, J. M., Kessler, J. L., Hyodo, M., Hayakawa, Y., & Lory, S. (2007a). A cyclic-di-GMP receptor required for bacterial exopolysaccharide production. *Mol. Microbiol., 65,* 1474-1484.

Lee, J., Jayaraman, A., & Wood, T. K. (2007b). Indole is an inter-species biofilm signal mediated by SdiA. *BMC Microbiol., 7,* 42.

Lee, J., Bansal, T., Jayaraman, A., Bentley, W. E., & Wood, T. K. (2007c). Enterohemorrhagic *Escherichia coli* biofilms are inhibited by 7-hydroxyindole and stimulated by isatin. *Appl. Environ. Microbiol., 73,* 4100-4109.

Lee, J., Maeda, T., Hong, S. H., & Wood, T. K. (2009). Reconfiguring the quorum-sensing regulator SdiA of *Escherichia coli* to control biofilm formation via indole and N-acetylhomoserine lactones. *Appl.Environ. Microbiol., 75,* 1703-1716.

Lemon, K. P., Earl A. M., Vlamakis, H. C., Aguilar, C., & Kolter, R. (2008). Biofilm development with an emphasis on *Bacillus subtilis. Curr. Top. Microbiol. Immunol., 322,* 1-16.

Lenard, J. (1992). Mammalian hormones in microbial cells. *Trends Biochem. Sci., 17,* 47-150.

Lyte, M. (1993). The role of microbial endocrinology in infectious disease. *J. Endocrinol., 137,* 343-345.

Lyte, M. (2004). Microbial endocrinology and infectious disease in the 21st century. *Trends Microbiol., 12,* 14-20.

Lyte, M. & Bailey, M. T. (1997). Neuroendocrine-bacterial interactions in a neurotoxin-induced model of trauma. *Journal of Surgical Research, 70,* 195-201.

Lyte, M. & Ernst, S. (1993). Alpha and beta adrenergetic receptor involvement in catecholamine-induced growth of gram-negative bacteria. *Biochem. Biophys. Res. Commun., 190,* 447-452.

Lyte, M., Frank, C. D., & Green, B. T. (1996). Production of an autoinducer of growth by norepinephrine-cultured *Escherichia coli* O157:H7. *FEMS Microbiol. Lett., 139,* 155-159.

Lyte, M., Erickson, A. K., Arulanandam, B. P., Frank, C. D., Crawford, M. A., & Francis, D. H. (1997a). Norepinephrine-induced expression of the K99 pilus adhesin of enterotoxigenic *Escherichia coli*. *Biochem. Biophys. Res. Commun., 27*, 682-686.

Lyte, M., Arulanandam, B. P., Nguyen, K. T., Frank, C., Erickson, A., & Francis, D. (1997b). Norepinephrine-induced growth and expression of virulence-associated factors in enterotoxigenic and enterohemorrhagic strains of *E. coli*. *Adv. Exp. Med. Biol., 412*, 331-339.

Lyte, M., Freestone, P. P. E., Neal, C. P., Olson, B. A., Haigh, R. D., Baystone, R., & Williams, P. H. (2003). Stimulation of *Staphylococcus epidermidis* growth and biofilm formation by catecholamine inotropes. *Lancet, 361*, 130-135.

Malikina, K.D., Shishov, V.A., Kudrin, V.S., & Oleskin, A.V. (2010). Regulatory role of monoamine neuromediators in *Saccharomyces cerevisiae* cells. *Appl. Biochem. Mikrobiol., 46*, in print.

Martins, F. S., Nardi, R. M. D., Arantes, R. M. E., Rosa, C. A., Neves, M. J., & Nicoli, J. R. (2005). Scanning of yeasts as probiotics based on capacities to colonize the gastrointestinal tract and to protect against enteropathogenic challenge in mice. *J. Gen. Appl. Microbiol., 51*, 83-92.

Masters, R. D. (2001). Biology and politics: linking nature and nurture. *Ann. Rev. Polit. Sci, 4*, 345-369.

Mittal, N., Budrene, E. O., Brenner, M. P., & Van Oudenaarden, A. (2003). Motility of *Escherichia coli* cells in clusters formed by chemotactic aggregation. *Proc. Natl. Acad. Sci. U.S.A., 100*, 13259-13263

Montal, M. (1998). Mitochondria, glutamate neurotoxicity, and the death cascade. *Biochim. Biophys. Acta., 1366*, 113-126.

Morgan, R., Kohn, S., Hwang, S. H., Hassett, D. J., & Sauer, K. (2006). BdlA, a chemotaxis regulator essential for biofilm dispersal in *Pseudomonas aeruginosa*. *J. Bacteriol., 188*, 7335-7343.

Murrison, R. (2001). Is there a role for psychology in ulcer disease? *Integrative Psychol. Behavioral Sci., 36*, 75-83.

Nikolaev, Yu. A. & Plakunov, V. K. (2007). A biofilm: A "city of microbes" or an analog of a multicellular organism? *Mikrobiologiya, 76*, 149-163.

Oleskin, A. V. (2007). Biopolitics. Political Potential of Modern Life Sciences. Moscow: Nauchnyi Mir.

Oleskin, A. V. & Kirovskaya, T. A. (2006). Research on population organization and communication in microorganisms. *Mikrobiologiya, 75*, 1-6.

Oleskin, A. V., Kirovskaya, T. A., Botvinko, I. V., & Lysak, L. V. (1998). Effects of serotonin (5-hydroxytryptamine) on the growth and differentiation of microorganisms. *Mikrobiologia, 67,* 251-257.

Pavlova, I. B., Levchenko, K. M., & Bannikova, D. A. (2007). Atlas Morfologii Populatsiy Patogennykh Bakteriy (Atlas of the Morphology of Populations of Pathogenic Bacteria). Moscow: Kolos.

Penfold, W. J. (1914). On the nature of the bacterial lag. *J. Hygiene., 14,* 215-241.

Rahman, M. A., Azuma, Y., Fukunaga, H., Murakam T., Sugi K., Fukushi, H., Miura, K., Suzuki, H., & Shizai, M. (2005). Serotonin and melatonin, neurohormones for homeostasis, as novel inhibitors of infection by the intracellular parasite chlamydia. *J. Antimicrob. Chemother., 56,* 861-868.

Rahn, O. (1904). Über den Einfluß der Stoffwechselprodukte auf das Wachstum der Bakterien. *Zbl. Bakteriol. Parasitenk., 16,* .417-429.

Romeo, T. (2006). When the party is over: a signal for dispersal of *Pseudomonas aeruginosa* biofilms. *J. Bacteriol., 188,* 7325-7327.

Roschina, V. V. (1991). Biomediatory v Rasteniyakh. Acetikholin i Biogennye Aminy (Biomediators in Plants. Acetylcholine and Biogenic Amines). Puschino: Puschino Bio-Center.

Ryan, R. P. & Dow, J. M. (2008). Diffusible signals and interspecies communication in bacteria. *Microbiology, 154,* 1845-1858.

Safronova I. Yu. & Botvinko, I. V. (1998). Extracellular matrix of *Bacillus subtilis* 271: polymer composition and functions. *Mikrobiologiya, 67,* 55-60.

Senadheera, D., & Cvitkovitch, D. G. (2008). Quorum sensing and biofilm formation by *Streptococcus mutans. Adv. Exp. Med. Biol., 631,* 178-188.

Shenderov, B. A. (2001). Meditsinskaya Mikrobnaya Ekologiya i Funktsional'noe Pitanie (Medical Microbial Ecology and Functional Nutrition). Moscow: Grant.

Shishov, V. A., Kirovskaya, T. A., Kudrin, V. S., & Oleskin, A. V. (2009). Amine neuromediators, their precursors, and oxidation products in the culture of *Escherichia coli* K-12. *Appl. Biochem. Mikrobiol., 45,* 494-497.

Shpakov, A. O. (2009). QS-type non-peptide signal molecules of bacteria. *Mikrobiologiya, 78,* 163-175.

Sircili, M. P., Walters, M., Trabulsi, L. R., & Sperandio, V. (2004). Modulation of enteropathogenic *Escherichia coli* virulence by quorum sensing. *Infect. Immun., 72,* 2329-2337.

Skulachev, V. P. (1999) Phenoptosis. Programmed organism death. *Biochemistry (Moscow), 64,* 1679-1688.

Somit, A. & Peterson, S. A. (1998). Biopolitics after three decades: a balance sheet. *Brit. J. Polit. Sci., 28,* 555-571.

Sperandio, V., Torrres, A. G., Jarvis, B., Nataro, J. P., & Kaper, J. (2003). Bacteria-host communication. The language of hormones. *Proc. Natl. Acad. Sci. USA., 100,* 8951-8956.

Stoodley, P., Sauer, K., Davies, D. G., & Costerton, J. W. (2002). Biofilms as complex differentiated communities. *Ann. Rev. Microbiol., 56,* 187-209.

Strakhovskaya, M. G., Ivanova, E. V., & Fraikin, G. Ya. (1993). Stimulatory effect of growth on growth of the yeast *Candida guillermondii* and the bacterium *Streptococcus faecalis. Mikrobiologiya, 62,* 46-49.

Sumina, E. (2006). Behavior of filamentous cyanobacteria in lab culture. *Mikrobiologiya, 75,* 532-537.

Sutherland, I. W. (2001). Biofilm polysaccharides: a strong and sticky framework. *Microbiology, 147,* 3-9.

Tecott, L. H. & Barondes, S. H. (1996). Behavioral genetics: genes and aggressiveness. *Curr. Biol., 6,* 238-240.

Tetz, V. V., Rybalchenko, O. V., & Savkova, G. A. (1993). Ultrastructure of the surface film. *J. Gen. Microbiol., 139,* 855-858.

Tetz, V. V., Korobov, V. P., Artemenko, N. K., Lemkina, L. M, Panjkova, N. V., & Tetz, G. V. (2004). Extracellular phospholipids of isolated bacterial communities. *Biofilms, 1,* 149-155.

Tsavkelova, E. A., Botvinko, I. V., Kudrin, V. S., & Oleskin, A. V. (2000). Detectsiya neiromediatornykh aminov u mikroorganizmov metodom vysokoeffektivnoy zhidkostnoy khromatografii (Detecting neuromediator amines in microorganisms with high performance liquid chromatography). *Dokl. Ross. Akad. Nauk, 372,* 840-842.

Valyshev, A. V. & Gilmutdinova, F. G. (2006). Microbial ecology of the human alimentary tract. In: O. V. Bukharin (Ed.), *Human Microbial Ecology.* (pp. 167-290). Ekaterinburg: Russian Academy of Sciences, Urals Division,.

Vlisidou, I., Lyte, M, van Diemen, P. M., Hawes, P., Monaghan, P., Wallis, T. S., & Stevens, M. P. (2004). The neuroendocrine stress hormone norepinephrine augments *Escherichia coli* O157:H7-induced enteritis and adherence in a bovine ligated ileal loop model of infection. *Infect. Immun., 72,* 5446-5451.

Walters, M. & Sperandio, V. (2006). Autoinducer 3 and epinephrine signaling in the kinetics of locus of enterocyte effacement gene expression in enterohemorrhagic *Escherichia coli. Infect. Immun.*, *74*, 5445-5455

Watnick, P. I., & Kolter, R., (2000). Biofilm, city of microbes. *J. Bacteriol.*, *182*, 2675-2679.

Whimpenny, J., Manz, W., & Szewzyk, U. (2000). Heterogeneity in biofilms. *FEMS Microbiol. Rev,. 24*, 661-671.

Xu, J., Bjursell, M. K., Himrod, J., Deng, S., Carmichael, L. K., Chiang, H. C., Hooper, L. V., & Gordon, J. I. (2003). A genomic view of the human-*Bacteroides thetaiomicron* symbiosis. *Science, 299*, 2074-2076.

Yildiz, F. H., Liu, X. S., Heydorn, A., & Schoolnik G. K.. (2004). Molecular analysis of rugosity in a Vibrio cholerae O1 El Tor phase variant. *Mol. Microbiol., 53,* 497-515.

Yurkevich, D. I. & Kutyshenko, V. P. (2002). The medusomycete (the tea fungus): history of research, composition, and physiological and metabolic peculiarities. *Biophysics (Moscow), 47,* 1116-1129.

Zaborina, O., Lepine, F., Xiao, G., Valuckaite, V., Chen, Y., Li, T., Ciancio, M., Zaborin, A., Petroff, E., Turner, J. R., Rahme, L. G., Chang, E., & Alverdy, J. C. (2007). Dynorphin activates quorum sensing quinolone signaling in *Pseudomonas aeruginosa. PloS Pathogens, 3,* e35.

Zumft, W. G. (1993). The biological role of nitric oxide in bacteria. *Arch. Microbiol., 160,* 253-264.

Reviewer:
Dr. V.P. Korobov
Ass. Prof., Ph.D.
Head of the Microorganisms' Development Biochemistry Laboratory
Institute of Ecology and Genetics of Microorganisms
Urals Branch of the Russian Academy of Sciences
Perm, Russia

Index

A

access, 30
acetic acid, 26
acid, 5, 8, 19, 23, 24, 26, 31, 33, 34, 35, 36, 37, 40
acidity, 15
adaptation, 8, 26
ADHD, 16
adhesion, 20, 26
agar, 21, 26
aggregates, 4, 5, 21
aggregation, vii, 21, 24, 26, 46
aggression, viii, 37
aggressive behavior, 19
aggressiveness, 48
agonist, 25, 28
alanine, 19, 33
alcohol, 38
alcohols, 38, 42
altruistic acts, 27
amines, vii, 1, 19, 26, 30, 33, 34, 35, 36, 37, 38, 48
amino acids, 19, 31, 33, 37
ammonium, 38
anger, 16
antibiotic, 16, 30
anxiety, 19
apoptosis, 29, 40
applications, 30
arginine, 29
ATP, 40
attachment, vii, 5, 6, 7, 25
autism, 16
axons, 23, 27

B

Bacillus subtilis, 34, 45, 47
bacteria, 2, 5, 7, 8, 11, 12, 16, 17, 20, 22, 24, 25, 26, 33, 34, 36, 38, 42, 43, 44, 45, 47, 49
bacterial infection, 28
bacterial strains, 30
bacterium, vii, 6, 11, 12, 20, 24, 28, 30, 34, 37, 40, 41, 43, 48
balance sheet, 48
behavior, viii, 2, 3, 19, 27, 29, 37
benzodiazepine, 41
bile, 20
binding, 22, 24, 25, 39
bioluminescence, 11
biomass, 20, 21, 24, 26, 27, 34, 35, 36, 37
biopolymer, vii, 5, 34
biosynthesis, 9, 35, 36, 37, 42
blocks, 8, 11, 22
blood, viii, 13, 20, 21, 29, 37
blood vessels, 29
blood-brain barrier, viii, 37
bloodstream, 20, 21, 23, 29, 34

bowel, 16, 34
brain, viii, 1, 19, 23, 27, 29, 37, 40
brain structure, 23
brainstem, 23

C

Ca^{2+}, 40
carbon, 19, 24, 37, 39
carbon monoxide, 19
caries, 1
catecholamines, 19, 20, 21, 22, 23, 24, 25, 26, 27, 28, 34, 35, 37, 38
cell, 1, 3, 4, 5, 6, 8, 20, 21, 24, 25, 26, 29, 31, 33, 35, 36, 43, 44
cell death, 29, 31
cell signaling, 43, 44
cell surface, 8
cellulose, 9
channels, 4
chemotaxis, 9, 41, 46
chlamydia, 24, 47
chlorine, 30
chromosome, 25
clusters, 32, 46
codes, 12
cohesion, 5
colitis, 16, 34
colon, 16, 34
colon cancer, 16
colonization, 8, 28, 41
color, 27
communication, 17, 21, 25, 26, 31, 39, 40, 43, 46, 47, 48
community, vii, 17
components, 6, 9, 11, 17, 34, 38
composition, 15, 16, 34, 47, 49
compounds, 1, 15, 21, 24, 25, 30, 35, 37
concentrates, 2, 39
concentration, 7, 9, 11, 12, 16, 24, 26, 28, 29, 30, 31, 40
Congress, iv
conjunctiva, 15
control, 8, 9, 28, 30, 42, 43, 45
conversion, 36

cooling, 3
coordination, 3
Copyright, iv
CSF, 36, 37
cultivation, 7, 31
cultivation conditions, 7
culture, vii, 22, 28, 34, 35, 36, 37, 39, 47, 48
cystic fibrosis, 4

D

damages, iv
death, 7, 31, 46, 48
degradation, vii, 35, 36
denitrification, 30
density, 11, 12, 17, 36, 43
dental caries, 4
derivatives, 25, 26
destruction, 4
diarrhea, 13
diet, 17
differentiation, 29, 31, 39, 47
diffusion, 34
digestion, 15
dihydroxyphenylalanine, 35
dilation, 29
diseases, 4, 16, 20
dispersion, vii
distribution, 3, 8, 30
diversity, 3, 4, 7, 17
division, 35
DNA, 5, 6
donors, 30
dopamine, vii, 1, 19, 20, 21, 22, 23, 24, 26, 34, 36, 37, 42
dopaminergic, 43
drugs, 27, 42
dynamics, 34, 36, 37

E

ecology, 48
ecosystem, 38
embryogenesis, 42

Index 53

emotional state, 29
employment, 16
encoding, 11, 31
endocarditis, 4
endocrinology, 39, 43, 45
endorphins, 19, 27
endothelial cells, 29
energy, 24
enkephalins, 19, 27
enteritis, 48
environment, 5, 8, 44
environmental factors, 5, 7, 9
environmental stimuli, 9
enzymes, 6, 35, 36, 37
epinephrine, 19, 20, 22, 23, 39, 41, 48
epithelial cells, 15, 25
epithelium, 15
eukaryote, 20, 21
eukaryotic cell, 22, 29, 40
euphoria, 27
evolution, 2
excitation, 23
exporter, 25, 44
exposure, 27

G

gases, 29
gastrointestinal tract, 44, 46
gene, 8, 11, 28, 29, 31, 39, 41, 43, 44, 48
gene expression, 41, 43, 49
genes, 7, 8, 11, 12, 22, 23, 25, 31, 48
genetic information, 8
genetics, 48
germination, 24
gift, 4, 6
gland, 1
glutamate, 9, 40, 46
glutamic acid, 33
glycine, 19
glycoproteins, 5
granules, 26
groups, 19, 21, 31, 33
growth, vii, 20, 21, 22, 23, 24, 26, 27, 28, 30, 34, 35, 36, 37, 38, 41, 42, 43, 44, 45, 46, 47, 48
growth rate, vii
gut, viii, 37

F

failure, 16
family, 12, 43
fatty acids, 15, 37
fear, 16
feces, 16, 20
females, 29
fibers, 8
films, 3
flight, 17
flora, 16, 44
fluid, vii, 22, 36, 37
food, 15, 20, 27, 30, 37
food poisoning, 37
fungi, 1, 42
fungus, 3, 34, 39, 49

H

habitat, 8
hallucinations, 23
harm, 4
health, 1, 17, 40
heat, 43
hepatocytes, 29
hexagonal lattice, 31
histamine, 1, 19, 26, 27, 30, 33
histidine, 26
homeostasis, 47
hormone, 16, 19, 24, 29, 39, 48
host, vii, 1, 5, 16, 17, 23, 26, 29, 31, 36, 38, 39, 40, 43, 48
hydrogen, 19, 30
hydrogen peroxide, 30
hydrophilicity, 7
hydrophobicity, 7
hypothalamus, 26

I

ileum, 21
immune response, 29
immune system, 15
immunoglobulin, 20
immunologist, 16
impulses, 2, 27, 34
in vitro, 8
incentives, 8
indigenous, 42
infection, 21, 23, 26, 27, 29, 47, 48
infectious disease, 17, 43, 44, 45
inflammation, 20, 23, 26, 27
inflammatory bowel disease, 34, 41
inflammatory disease, 20
information processing, 40
inguinal, 15
inhibition, 45
injury, iv, 3
inoculation, 21
insulin, 39
integrity, 5, 44
interaction, 22
interactions, 5, 21, 26, 40, 45
intestine, 44
ions, 40

K

killing, 41
kinetics, 48

L

labor, 8
landscape, 26
language, 48
large intestine, 4, 15, 36
learning, 42
life cycle, 21
life sciences, 1
lifestyle, 6, 8
likelihood, 7
limitation, 7, 38
links, 28
lipids, 7
liquid chromatography, vii, 34, 48
liver, 16
locus, 48
LSD, 23
lumen, vii, 15, 20, 23, 27, 36
lysergic acid diethylamide, 23

M

macrophages, 29
maintenance, 15
majority, vii, 3, 5, 34
malnutrition, 16
maltose, 37
mammal, 34
masking, 5
matrix, vii, 2, 3, 5, 6, 7, 8, 9, 24, 34, 47
maturation, 15
medium composition, 7
melatonin, 24, 26, 47
mental disorder, 16
mental health, 16
metabolism, 39
metabolites, vii, 5
mice, 20, 27, 29, 46
microbial cells, vii, 5, 6, 7, 8, 11, 12, 21, 30, 35, 40, 45
microorganism, 37
microstructure, 31
middle ear infection, 4
milk, 17
mitochondria, 40
model, 45, 48
molecular weight, 34
molecules, 5, 11, 13, 34, 36, 38, 39, 40, 47
monolayer, 7
monomers, 8
mood, 40
morphine, 28
morphogenesis, 42
Moscow, 42, 43, 45, 46, 47, 48, 49
mothers, 16

Index

mucosa, 15, 20, 26, 27, 36, 42, 44
mucous membrane, vii, 4
mucus, 7, 15
multicellular organisms, 3
mutant, 12, 30, 31
mutation, 29

N

negative consequences, 27
nerve, 27
nervous system, 13, 23, 27, 34, 37, 39, 40
network, 9
neurodegenerative diseases, 40
neurons, 1, 22
neuropeptides, 27
neurophysiology, 2
neurotoxicity, 46
neurotransmitter, 1, 13, 19, 23, 26, 29, 31, 33, 34, 35, 36, 37, 38, 39
nitric oxide, vii, 9, 19, 29, 30, 31, 33, 41, 44, 49
nitrogen, 24, 38
norepinephrine, vii, 1, 19, 20, 21, 22, 23, 26, 27, 34, 35, 36, 37, 41, 43, 45, 48
nuclei, 23
nucleotides, 37

O

operon, 11
opioids, 27
oral cavity, 6
order, 15
organ, 4, 11, 16
organism, vii, viii, 1, 2, 4, 6, 15, 17, 20, 23, 28, 29, 30, 31, 37, 38, 39, 46, 48
osteomyelitis, 4
oxidation, 36, 37, 47
oxidation products, 36, 37, 47
oxygen, 9

P

pain, 27, 34
pancreas, 39
parasite, 47
parasitic infection, 44
pathogens, vii, 16, 20, 23, 24, 26
penicillin, 34
peptides, vii, 6, 19
performance, 34, 48
permission, iv
personal communication, 23, 25
pertussis, 20
phenylalanine, 38
phospholipids, 48
phosphorylation, 22
pineal gland, 24
plants, 1, 24
pleasure, 19
pneumonia, 22
politics, 46
pollutants, 16
polymer, 47
population, viii, 11, 15, 16, 28, 29, 38, 46
population density, 11
positive reinforcement, 27
predation, 8
probiotic, 20
production, 8, 9, 25, 27, 29, 33, 44, 45
prokaryotes, 30, 33
proliferation, 20, 21, 24, 26, 29, 31, 35, 36
promoter, 12, 43
properties, 4, 12, 26
propranolol, 22
proteins, 5, 6, 12, 22, 25, 31, 37
Pseudomonas aeruginosa, 6, 20, 41, 43, 46, 47, 49
psychology, 46

R

radioactive waste, 16
range, 3, 34, 36
reactive oxygen, 29

receptor sites, 22
receptors, 21, 22, 24, 27, 36, 39, 40
recognition, 5
recommendations, iv
regenerate, 3
regulation, vii, 9, 15
regulators, 8, 22, 42, 43
relevance, 1
reproduction, 28
resources, 8
respect, viii, 8, 15, 31, 39, 40
respiratory, 20
rheumatoid arthritis, 16
rights, iv
RNA, 6, 12
Russia, 49

S

scanning electron microscopy, 20
searching, 8
seeding, 7
sensing, 6, 8, 11, 12, 13, 17, 21, 22, 25, 28, 38, 42, 43, 44, 45, 47, 49
sensitivity, 30, 34
separation, 16
septic shock, 29
serotonin, vii, 1, 19, 23, 24, 25, 26, 27, 30, 33, 34, 35, 36, 37, 47
serum, 21
severe stress, 17
sewage, 3
shear, 7
signal transduction, 12
signaling pathway, 42
signals, vii, 1, 9, 23, 25, 34, 38, 47
skin, 15, 20, 38
social behavior, 1, 2, 37, 39, 44
social life, 8
social stress, 16
sodium, 12, 29, 30
sodium dodecyl sulfate (SDS), 12
soil, 24
space, 17

species, vii, 1, 3, 5, 7, 8, 9, 12, 15, 16, 20, 23, 29, 30, 31, 42, 45
stimulus, 23
stimulus perception, 23
strain, 7, 12, 13, 20, 21, 22, 23, 24, 26, 27, 30, 31, 43
strategies, 30
streptococci, 7
stress, 7, 16, 17, 20, 26, 28, 29, 34, 40, 48
stress factors, 28
structuring, vii
sucrose, 38
sulfur, 19
suppression, 6, 28
surfactant, 12
suspensions, 8
symbiosis, 41, 49
synaptic clefts, 23
syndrome, 16, 34
synthesis, vii, 6, 8, 9, 11, 20, 24, 25, 27, 28, 34, 36

T

temperature, 7, 15
terminals, 27
thinking, 40
threat, 4
threshold, 11, 12, 26
threshold level, 11, 12
thyroid, 39
thyroxin, 39
tissue, 1
toxic effect, 29
transcription, 9, 12, 22
transducer, 31
transition, 6, 7, 30, 35, 38
translation, 9
transmission, 27, 34
trauma, 45
tryptophan, 23, 38
tumor, 29
tumor cells, 29
tyrosine, 19

U

ulcer, 17, 46
upper respiratory tract, 15
USSR, 45

V

variables, 41
viruses, 15

vitamins, 15, 37

W

weapons, 29
wood, 3

Y

yeast, 4, 16, 30, 34, 37, 38, 48